# Shengmai San

## Traditional Herbal Medicines for Modern Times

Each volume in this series provides academia, health sciences and the herbal medicines industry with in-depth coverage of the herbal remedies for infectious diseases, certain medical conditions or the plant medicines of a particular country.

Edited by Dr Roland Hardman

**Volume 1**
*Shengmai San*, edited by Kam-Ming Ko

# Shengmai San

**Kam-Ming Ko**, Hong Kong University of Science & Technology, Kowloon, Hong Kong, China

London and New York

First published 2002 by Taylor & Francis
11 New Fetter Lane, London EC4P 4EE

Simultaneously published in the USA and Canada
by Taylor & Francis Inc,
29 West 35th Street, New York, NY 10001

*Taylor & Francis is an imprint of the Taylor & Francis Group*



Typeset in 11/12pt Garamond 3 by
Graphicraft Limited, Hong Kong
Printed and bound in Great Britain by
TJ International Ltd, Padstow, Cornwall



*British Library Cataloguing in Publication Data*
A catalogue record for this book is available
from the British Library

*Library of Congress Cataloging in Publication Data*
A catalog record has been requested

ISBN 0-415-28490-2

# Contents

# Contributors

**Bao-Jing Lu**
Department of Cardiology
Xin-Hua Hospital
Shanghai Second Medical University
1665 Kong-Jiang Road
Shanghai 200092
China

**Chun-Tao Che**
School of Chinese Medicine
The Chinese University of Hong Kong
Shatin, N.T.
Hong Kong
China

**Duncan H.F. Mak**
Department of Biochemistry
Hong Kong University of Science
& Technology
Clear Water Bay
Hong Kong
China

**Jie Chen**
Department of Cardiology
Xin-Hua Hospital
Shanghai Second Medical University
1665 Kong-Jiang Road
Shanghai 200092
China

**Kam-Ming Ko**
Department of Biochemistry
Hong Kong University of Science
& Technology
Clear Water Bay
Hong Kong
China

**Liang-Yuan Zheng**
Livzon Pharmaceutical Group, Inc.
North Guihua Road, Gongbei
Zhuhai, Guangdong 519020
China

**Mei-Hua Zhao**
Department of Cardiology
Xin-Hua Hospital
Shanghai Second Medical University
1665 Kong-Jiang Road
Shanghai 200092
China

**Shang-Biao Lu**
Department of Cardiology
Xin-Hua Hospital
Shanghai Second Medical University
1665 Kong-Jiang Road
Shanghai 200092
China

**Shu-Ying Chen**
School of Materia Medica
Guangzhou University of Traditional
Chinese Medicine
12 Jichang Road, Guangzhou
519020
China

**Si-Qian Liang**
Department of Applied Science
Hong Kong Institute of Vocational
Education (CW)
Chai Wan
Hong Kong
China

**Siu-Po Ip**
School of Chinese Medicine
The Chinese University of Hong Kong
Shatin, N.T.
Hong Kong
China

**Song-Ming Liang**
School of Chinese Medicine
The Chinese University of Hong Kong
Shatin, N.T.
Hong Kong
China

**Timothy C.M. Tam**
School of Pharmacy
The Chinese University of Hong Kong
Shatin, N.T.
Hong Kong
China

**Tze-Kin Yim**
Department of Biochemistry
Hong Kong University of Science & Technology
Clear Water Bay
Hong Kong
China

**Xiang-Yang Zhu**
Department of Cardiology
Xin-Hua Hospital
Shanghai Second Medical University
1665 Kong-Jiang Road
Shanghai 200092
China

**Ya-Chen Zhang**
Department of Cardiology
Xin-Hua Hospital
Shanghai Second Medical University
1665 Kong-Jiang Road
Shanghai 200092
China

**Ye-Zhi Rong**
Dept. of Medicine, Xin-Hua Hospital
Shanghai Second Medical University
1665 Kong-Jiang Road
Shanghai 200092
China

**Yong-Qing Yan**
The China Pharmaceutical University
24 Tong Jia Xiang, Nanjin
Jiangsu 210009
China

**Zhi-Min Xu**
Department of Cardiology
Xin-Hua Hospital
Shanghai Second Medical University
1665 Kong-Jiang Road
Shanghai 200092
China

# Preface to the series

Global warming and global travel are among the factors resulting in the spread of such infectious diseases as malaria, tuberculosis, hepatitis B and HIV. All these are not well controlled by the present drug regimes. Antibiotics too are failing because of bacterial resistance. Formerly less well known tropical diseases are reaching new shores. A whole range of illnesses, for example cancer, occur worldwide. Advances in molecular biology, including methods of *in vitro* testing for a required medical activity give new opportunities to draw judiciously upon the use and research of traditional herbal remedies from around the world. The re-examining of the herbal medicines must be done in a multidisciplinary manner.

Since 1997 twenty volumes have been published in the Book Series **Medicinal and Aromatic Plants – Industrial Profiles**. The series continues, and is characterised by a single plant genus per volume. With the same Series Editor, this new series **Traditional Herbal Medicines for Modern Times**, covers multi genera per volume. It accommodates for example, the Traditional Chinese Medicines (TCM), the Japanese Kampo versions of this and the Ayurvedic formulations of India. Collections of plants are also brought together because they have been re-evaluated for the treatment of specific diseases, such as malaria, tuberculosis, cancer, diabetes, etc. Yet other collections are of the most recent investigations of the endemic medicinal plants of a particular country, e.g. of India, South Africa, Mexico, Brazil (with its vast flora), or of Malaysia with its rainforests said to be the oldest in the world.

Each volume reports on the latest developments and discusses key topics relevant to interdisciplinary health science research by ethnobiologists, taxonomists, conservationists, agronomists, chemists, pharmacologists, clinicians and toxicologists. The Series is relevant to all these scientists and will enable them to guide business, government agencies and commerce in the complexities of these matters. The background to the subject is outlined below.

Over many centuries, the safety and limitations of herbal medicines have been established by their empirical use by the 'healers' who also took a holistic approach. The 'healers' are aware of the infrequent adverse affects and know how to correct these when they occur. Consequently and ideally, the pre-clinical and clinical studies of a herbal medicine need to be carried out with the full cooperation of the traditional healer. The plant composition of the medicine, the stage of the development of the plant material, when it is to be collected from the wild or when from cultivation, its post-harvest treatment, the preparation of the medicine, the dosage and frequency and much other essential information is required. A consideration of the intellectual property rights and appropriate models of benefit sharing may also be necessary.

Wherever the medicine is being prepared, the first requirement is a well documented reference collection of dried plant material. Such collections are encouraged by organisations like the World Health Organisation and the United Nations Industrial Development Organisation. The Royal Botanic Gardens at Kew in the UK is building up its collection of traditional Chinese dried plant material relevant to its purchase and use by those who sell or prescribe TCM in the UK.

In any country, the control of the quality of plant raw material, of its efficacy and of its safety in use, are essential. The work requires sophisticated laboratory equipment and highly trained personnel. This kind of 'control' cannot be applied to the locally produced herbal medicines in the rural areas of many countries, on which millions of people depend. Local traditional knowledge of the 'healers' has to suffice.

Conservation and protection of plant habitats is required and breeding for biological diversity is important. Gene systems are being studied for medicinal exploitation. There can never be too many seed conservation 'banks' to conserve genetic diversity. Unfortunately such banks are usually dominated by agricultural and horticultural crops with little space for medicinal plants. Developments such as random amplified polymorphic DNA enable the genetic variability of a species to be checked. This can be helpful in deciding whether specimens of close genetic similarity warrant storage.

From ancient times, a great deal of information concerning diagnosis and the use of traditional herbal medicines has been documented in the scripts of China, India and elsewhere. Today, modern formulations of these medicines exist in the form of e.g. powders, granules, capsules and tablets. They are prepared in various institutions e.g. government hospitals in China and Korea, and by companies such as Tsumura Co. of Japan with good quality control. Similarly, products are produced by many other companies in India, the USA and elsewhere with a varying degree of quality control. In the USA, the dietary supplement and Health Education Act of 1994 recognised the class of physiotherapeutic agents derived from medicinal and aromatic plants. Furthermore, under public pressure, the US Congress set up an Office of Alternative Medicine and this office in 1994 assisted the filing of several Investigational New Drug (IND) applications, required for clinical trials of some Chinese herbal preparations. The significance of these applications was that each Chinese preparation involved several plants and yet was handled as a *single* IND. A demonstration of the contribution to efficacy, of *each* ingredient of *each* plant, was not required. This was a major step forward towards more sensible regulations with regard to phytomedicines.

Something of the subject of western herbal medicines is now being taught again to medical students in Germany and Canada. Throughout Europe, the USA, Australia and other countries, pharmacy and health related schools are increasingly offering training in phytotherapy. TCM clinics are now common outside of China, and an Ayurvedic Hospital now exists in London and a degree course in Ayurveda is also available here.

The term 'integrated medicine' is now being used which selectively combines traditional herbal medicine with 'modern medicine'. In Germany there is now a hospital in which TCM is integrated with western medicine. Such co-medication has become common in China, Japan, India, and North America by those educated in both systems. Benefits claimed include improved efficacy, reduction in toxicity and the period of medication, as well as a reduction in the cost of the treatment. New terms such as adjunct therapy, supportive therapy and supplementary medicine now appear as a consequence of such co-medication. Either medicine may be described as an adjunct to the other depending on the communicator's view.

Great caution is necessary when traditional herbal medicines are used by those doctors not trained in their use and likewise when modern medicines are used by traditional herbal doctors. Possible dangers from drug interactions need to be stressed.

Dr Roland Hardman
January 2002

# Preface

The practice of traditional Chinese medicine (TCM) commonly prescribes herbal formula for the prevention and treatment of diseases. Shengmai San (SMS), a famous Chinese medicinal formula that has been used for more than eight hundred years in China, is comprised of *Radix Ginseng*, *Fructus Schisandrae* and *Radix Ophiopogonis*. Traditionally, SMS is used for the treatment of excessive loss of *essence Qi* and *body fluid* that threaten heart failure, particularly in summer when heat exhaustion and profuse sweating commonly occur. SMS, which can restore blood volume and prevent myocardial infarction, is also prescribed contemporarily for patients with coronary heart disease and various cardiovascular disorders. With the support from a number of experts involved in different areas of TCM, particularly the experimental and clinical research on SMS, this book was compiled to provide a comprehensive treatise on the historical, phytochemical, pharmacological/toxicological, clinical as well as pharmaceutical aspect of SMS and its component herbs. This monograph therefore provides a scientific rationale of using multi-component formulation in TCM for the prevention and treatment of diseases.

Kam-Ming Ko, Ph.D.
February 2001

# 1 Shengmai San – a renowned traditional Chinese medicinal formula

*Song-Ming Liang, Shu-Ying Chen, Si-Qian Liang* [Translated by Kam-Ming Ko]

## THE THEORY OF TRADITIONAL CHINESE MEDICINE IN PRESCRIPTION

### I. Treatment by differentiation of signs and symptoms

The basis of disease treatment in traditional Chinese medicine (TCM) emphasizes the principle of 'treatment by differentiation of signs and symptoms', which requires the identification of the etiology (causes of the symptoms), the definition of the therapeutic principles, the prescription of medicinal formula and the choice of herbs.

The *Differentiation of Signs and Symptoms* is a know-how, based on various stages of development of the disease, to analyze and diagnose in terms of syndromes. In other words, it is a process in which the cause, pathogenesis, nature and location of a disease is defined and served as the basis for 'treatment'.

### II. The eight therapeutic principles

'Treatment' is a process of applying therapeutic principles and herbal prescriptions to rectify the abnormal body condition, based on information derived from the 'differentiation of symptoms'. Under the theory of TCM there are *Eight Therapeutic Principles*, namely; *diaphoresis*, *inducing emesis*, *purgation*, *reconciliation*, *warming*, *heat-clearing*, *resolution* and *invigoration*, each of which targets the *Eight Principal Syndromes*: *Yin* and *Yang*, *superficies* and *interior*, *cold* and *heat*, *deficiency (asthenia)* and *sthenia*. *Diaphoresis* is a treatment for expelling superficially located evils by opening pores of the skin, regulating the function of *Ying* and *Wei*, and inducing perspiration. *Emetic* therapy is a treatment for the elimination of harmful substances from the throat, esophagus and stomach by the application of drug or physical stimuli that can induce vomiting. *Purgation* is a treatment for eliminating undigested food, *sthenic heat-evil* and *fluid* by the application of purgatives. *Reconciliation* is a treatment for regulating the *cold* and *heat* of the *superficies* and *interior*, as well as coordinating the functions of the *viscerae* and *organs*. *Warming* is a treatment for *cold* syndrome with medicine of a *warm* and *hot* nature. *Heat-clearing* therapy is a treatment for clearing away the *heat-evil* with *cold*-natured drugs. *Dispelling* therapy is a treatment for dissipating *sthenic evils*, such as stagnated energy, *blood stasis*, *phlegm-wetness* and indigestion. *Invigoration* is treatment for various types of deficiency-syndrome due to insufficiency of the *Yin*, *Yang*, *Qi* and *Blood*.

The *Eight Therapeutic Principles* have long been documented in the *Canon of Internal Medicine*. However, it is not until the *Qing* dynasty that Cheng (1732) provided a systematic evaluation and summary of the principles in his '*Summary on Medicine from*

*Clinical Practice*', which states that 'diagnosis of diseases can be generalized into *Yin* and *Yang*, *superficies* and *interior*, *cold* and *heat*, *deficiency (asthenia)* and *sthenia*, while treatment can be fully effected by *diaphoresis*, *inducing emesis, purgation, reconciliation*, *warming*, *heat-clearing*, *resolution* and *invigoration*.' The *Eight Therapeutic Principles* should therefore be applied on the ground of the *Eight Principal Syndromes*.

Although the *Eight Therapeutic Principles* have been tailored-made for the *Eight Principal Syndromes*, clinical manifestation of diseases is usually complex and not readily tackled by a single therapeutic method. A combination of a few therapeutic interventions could assure a complete remedy for the condition. In fact, the effect of treatment could be varied further by the priority as well as the extent of applying one or more of the *Eight Therapeutic Principles*. Therefore, the clinical application of these therapeutic methods should be based on the syndromes and the course of disease development before coming up with the appropriate prescriptions.

The theory of TCM holds that treatment of diseases should be guided by the theory and therapeutic principles, and implemented by applying formulas and herbs (or drugs). This bespeaks much of the importance of the theoretical basis and therapeutic principles for choosing the right formula and herbs for treatment. In TCM, the use of multi-component prescriptions is far more common than single herb in the prevention and treatment of diseases. Prescriptions are formulated on the basis of *Differentiation of Signs and Symptoms*, and the subsequent application of therapeutic principles and the right choice of drugs. Through an understanding of individual drug properties and their logical combinations, the therapeutic potential of the drugs could be maximized, with their toxic side effects being minimized or neutralized. The synergistic action exhibited by multi-component prescriptions is therefore effective for the management of more complicated diseases.

## III. The monarch, minister, assistant and guide of herbal formulas

Formulating a prescription requires, but is not limited to, the understanding of the disease progress and the principle of *Treatment by Differentiation of Signs and Symptoms*. A prescription is not a random mix of individual herbs nor a casual addition of therapeutic effect from individual component. The formulation is governed by a rationale under which herbs are assigned as *Monarch*, *Minister*, *Assistant* or *Guide*, representing their principal therapeutic roles and their inter-relationships. Based on the description by various practitioners a summary of such classification is given as follows:

The *Monarch* refers to the component herb(s) in a prescription that possess(es) the principal therapeutic action against the core disease or the major symptoms. It is the most important herb in a prescription that possesses the most potent pharmacological action and constitutes a greater proportion than that of the *Minister* or *Assistant*.

The *Minister* usually constitutes a lesser proportion and is less potent than the *Monarch*. It may be refered to as the component that enhances the action of the *Monarch* in treating severe diseases or symptoms. Or it may be referred to as the component that is targeted at treating complications or minor symptoms associated with a disease.

The *Assistant*, in turn, is less potent and constitutes a lesser proportion than the *Minister*. The *Assistant* may serve in three roles: (1) to enhance the actions of the *Monarch* and *Minister* or to treat minor symptoms directly; (2) to relieve or counteract the toxicity and side effects of the *Monarch* and *Minister*; (3) under some pathological conditions, to produce a synergistic effect with the *Monarch* even though it may possess opposing *nature* and *taste* as compared with the *Monarch*.

The *Guide* is a mild drug that is used in small amounts in a prescription. It may be used for facilitating the delivery of the various components in the formula to target organs, or to regulate the activities of other herbs.

When composing a prescription, the number of herbs assuming the role of *Monarch* should best be limited to unity, or at most two under more complicated situations, since more *Monarchs* would contribute to reduced efficacy due to dispersed pharmacological actions and undesirable pharmacological interactions. On the other hand, the number of *Ministers* could be larger than that of the *Monarch*, while the number of *Assistants* could be larger still, but that for *Guides* should be one or two. After all, the composition of a prescription should be formed with reference to the progress of the disease and the choice of *therapeutic principles*, as well as the pharmacological properties of the herbs. The concept of *Monarch*, *Minister*, *Assistant* and *Guide* is further exemplified by the *Herba Ephedrae* Decoction, as illustrated below.

## IV. Example of a TCM prescription

The *Herba Ephedrae* Decoction is indicated for the treatment of cold and flu caused by exogenous evils. These evils arise from *wind-cold*, with symptoms like: intolerance to cold weather, fever, headache, body pain, shortness of breath without sweating, thin and white fur on the tongue, floating and tense pulse. The pathogenesis of the disease involves the invasion of pathogens that block the *superficies* and trigger the body's immune response, as manifested by fever, intolerance to cold, headache and body pain. According to the theory of TCM, the *lungs* control the *vital energy* (*Qi*) of the body and are functionally coupled with the skin. Pathogenic changes such as blockade of the skin and obstruction of the pulmonary-*Qi* therefore give rise to explain the symptoms of the shortness of breath without sweating. Therefore, treatment should be aimed at inducing perspiration (*diaphoresis*), relieving *superficies*, clearing pulmonary obstruction and suppressing shortness of breath. The prescription includes *Herba Ephedrae* 9 g, *Ramulus Cinnamomi* 6 g, *Semen Armeniacae* 6 g and *Radix Glycyrrhizae* 3 g. Components of Herba Ephedrae Decoction are:

*Herba Ephedrae* – the *Monarch*: inducing perspiration and diaphoretic to disperse *wind-cold*; releasing pulmonary-*Qi* to suppress shortness of breath.

*Ramulus Cinnamomi* – the *Minister*: inducing perspiration, relieving muscular stiffness, *warming meridians* to activate *Yang*; capable of enhancing the diaphoretic action of *Herba Ephedrae*, as well as regulating *Ying-Qi* and *Wei-Qi*, relieving pains in the body and appendages.

*Semen Armeniacae* – the *Assistant*: releasing and suppressing pulmonary-*Qi* to assist *Herba Ephedrae* in suppressing shortness of breath; dispersing *wind* and *cold* to enhance the diaphoretic actions of *Herba Ephedrae* and *Ramulus Cinnamomi*.

*Radix Glycyrrhizae* – the *Guide*: regulating and neutralizing various herbs; to neutralize the *pungent* and *dry* properties of *Ramulus Cinnamomi* and prevent excessive perspiration induced by *Herba Ephedrae*.

It should be noted, however, that not every prescription carries *Minister*, *Assistant* and *Guide* and a single herb may not have the same role in every prescription. If the disease or condition is simple, treatment with a single or two herbs will suffice. If neither the *Monarch* nor *Minister* is toxic or too potent, there is no need for an *Assistant*. On the other hand, a *Guide* would not be needed should the main drugs be able to reach the target organs by themselves. For example, Shengmai San, which is indicated for the deficiency of *Yin* and *Qi*, comprises three herbs: *Radix Ginseng* serving as the *Monarch*, *Radix Ophiopogonis* as the *Minister* and *Fructus Schisandrae* as the *Assistant*.

The application of the theory encompassing *Monarch*, *Minister*, *Assistant* and *Guide* in prescriptions was a milestone in the development of disease treatment in TCM and a culmination of a huge amount of clinical experience over time.

## THE ORIGIN OF SHENGMAI SAN – THE RATIONALE OF FORMULATION

Shengmai San (SMS) is a well-known TCM formula that has been thought to originate from the *Neiwaishang Bianhuolun* (*Differentiation on Endogenous and Exogenous Diseases*) (Li 1213), but was later validated to originate from *Yixue Qiyuan* (*Origin of Medicine*) (Anonymous 1186). It is a formula that employs the therapeutic principles of invigorating the *Qi* and promotes *body fluid* production, and is indicated for symptoms of *heat*-induced depletion in *primordial-Qi*, *Qi* and *body fluid*, or deficiencies of the *Yin* and *Qi*. It comprises the three herbs, namely, *Radix Ginseng*, *Radix Ophiopogonis* and *Fructus Schisandrae*.

SMS is indicated for impairment in the regulation of the *Qi* and *body fluid* as manifested by profuse sweating, lassitude, shortness of breath, reluctance of speech, dryness of the larynx, thirst, asthenic and rapid pulse, as well as deficiency of the *lung* due to prolonged coughing. It is also indicated for deficiencies of the *Yin* and *Qi*, as manifested by dry cough with little phlegm, shortness of breath, spontaneous sweating, dryness of the mouth and tongue, asthenic and small pulse.

According to the theory of TCM, sweat is part of the *body fluid* that originates from the *heart*. Profuse sweating would thus impair the *heart-Yin*, as manifested by dryness of the mouth and tongue, dysphoria, thirst, weak and asthenic pulse. On the other hand, the *lung* controls the *Qi*. Spontaneous sweating would thereby consume the *Qi* and cause impairment in the *lung*, as manifested by shortness of breath and lassitude. Patients with impairment of the *lung* by prolonged coughing usually exhibit deficiencies of both the *Yin* and *Qi*, and therefore can be treated by invigorating the *Qi*, nourishing the *Yin* and promoting the production of the *body fluid*.

*Radix Ginseng* assumes the role of the *Monarch* in the SMS prescription. It is *sweet* in *taste* and *warm* in *nature*, capable of supplementing the *lung* with its actions of invigorating the *Qi* and promoting *body fluid* production. *Radix Ophiopogonis* assumes the role of the *Minister* by nourishing the *lung* and promoting *body fluid* production. It is *sweet* in *taste* and *cold* in *nature*, capable of nourishing *Yin* and clearing away *heat*. The combined use of *Radix Ginseng* and *Radix Ophiopogonis* produces a synergistic effect of invigorating the *Qi* and nourishing the *Yin*. *Fructus Schisandrae*, which is *sour* in *taste* and *warm* in *nature*, assumes dual roles of the *Assistant* and *Guide* in SMS. It is capable of astringing the *lung*, stopping excessive sweating and thirst, and promoting *body fluid* production. When the supplementing, nourishing and astringing functions of the three herbs are put together, they produce the effect of invigorating the *Qi*, nourishing the *Yin*, promoting *body fluid* production, quenching thirst, astringing the *Yin* and stopping excessive sweating. SMS, literally means 'decoction for restoring pulse', and was coined by its main action of invigorating the *Qi*, which is vital for sustaining a strong pulse. In addition, SMS may also be used to treat the deficiencies the *Yin* and *Qi* associated with deficiency of the *lung* induced by prolonged coughing.

With hundreds of years of practical experience, the efficacy of SMS has been confirmed and widely adopted in clinical situations. It has attracted much attention from the medical field and a large amount of research and clinical trials have been initiated since the 1970s. Clinical application of SMS in the treatment of cardiovascular diseases has been increasing, including the treatment of heart failure, ischemic cardiomyopathy and shock.

## CHEMICAL AND PHARMACOLOGICAL STUDIES ON THE FORMULATING PRINCIPLES OF SMS

The China Academy of Traditional Chinese Medicine have compared the effects of individual herbs and a standard preparation of SMS injection on the coronary flow and myocardial contractility of rat hearts, using Langendorff perfusion technique. Using pituitrin-(4 U/L)induced global ischemia in isolated perfused rat hearts, the effects of herbal preparations from various combinations of component herbs of SMS (7 combinations in total) were examined. Zhu *et al.* (1998a) have investigated the dynamic changes in chemical composition in SMS preparations and found that decoction of *Radix Ophiopogonis* and *Fructus Schisandrae* mixture could generate the antioxidant, 5-hydroxymethyl-2-furaldehyde (5-HMF), the formation of which increased as the amount of *Radix Ophiopogonis* was increased. In another study by the same authors (Zhu *et al.* 1998b), it was found that the saponin contents in various combinations of the component herbs varied, and the major saponin component in SMS were $Rg_2$, $Rg_3$, $Rh_1$ and $Rg_f$. While the content of $Rg_3$ and $Rh_1$ were increased in SMS, the contents of Rb, Rc, Rd and Re were decreased beyond detection. With regard to the $Rg_3$ and $Rh_1$, the optimal ratio of *Radix Ginseng*, *Radix Ophiopogonis* and *Fructus Schisandrae* should be 1:3:1.5.

### I. Effects of SMS and its individual herbal component in isolated normal rat heart

#### *A. Effects on coronary flow (Figure 1.1)*

At a concentration of 1.0 g/L, *Radix Ginseng* and *Fructus Schisandrae* extracts produced a similar effect in terms of time course and extent in enhancing coronary flow, with the effect peaking at 5–10 minutes after drug administration and then declined gradually. While *Radix Ophiopogonis* extract alone did not produce any enhancing effect on coronary flow, SMS produced a slow-onset, sustained and potent action that peaked at 20 minutes. The effect of SMS observable at 30 minutes was larger than that of the combination of the three individual herbs administered at the same time.

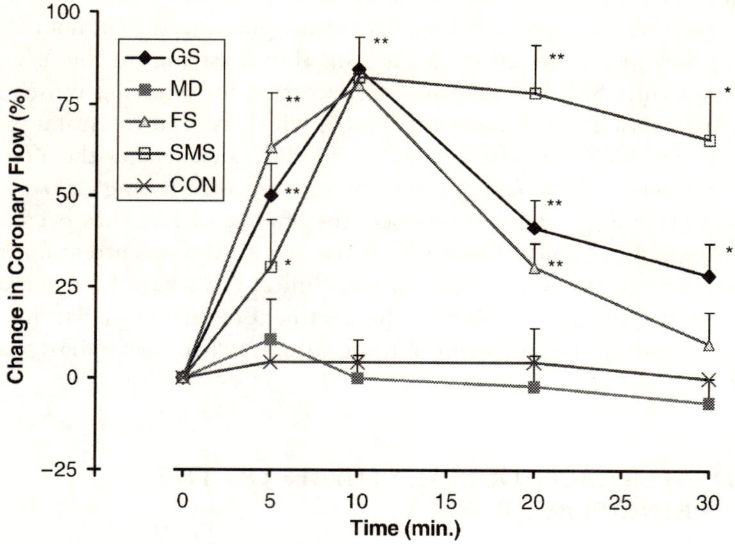


*Figure 1.1* Effect of SMS and extracts of its component herbs on coronary flow of normal rats. (All extracts were administered at 1.0 g/l) Herbal extracts: GS – *Radix Ginseng*; MD – *Radix Ophiopogonis*; FS – *Fructus Schisandrae*; SMS – Shengmai San; CON – Control.

*, ** $p < 0.05$ and $p < 0.01$, respectively, when compared with the control.

### *B. Effects on myocardial contractility (Figure 1.2)*

SMS and the three component herbs showed a tendency to enhance myocardial contractility at low concentrations and inhibit myocardial contraction at higher concentrations ($p < 0.01$), except for *Fructus Schisandrae* extract which showed a potent inotropic effect ($p < 0.05$) at all doses.

## II. Study of the protective effects of SMS, its component herbs and combinations against myocardial ischemia in isolated rat hearts

Dosage of extracts: *Radix Ginseng* 0.1 g/l, *Radix Ophiopogonis* 0.3 g/l, *Fructus Schisandrae* 0.15 g/l, SMS: sum of individual herbal component.

### *A. Effect on coronary flow (Figure 1.3)*

The three extracts of individual component herbs enhanced the coronary flow in ischemic rat hearts to varying extents. *Radix Ginseng* showed a slow onset of action, with a gradual increase in activity over time that surpassed those of *Radix Ophiopogonis* and *Fructus Schisandrae* beyond 15 min. *Radix Ophiopogonis*, on the other hand, produced a rapid onset but short duration of action, while *Fructus Schisandrae* exhibited only a mild effect. SMS extract produced a similar action to *Radix Ginseng* but with a higher potency; it could significantly reverse coronary spasm induced by pituitrin and increased coronary flow by 71.8 per cent at 30 minutes after drug administration ($p < 0.01$).

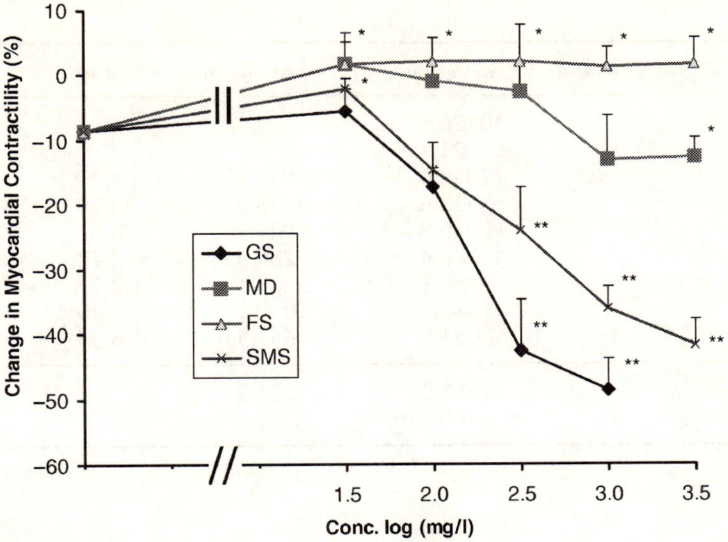


*Figure 1.2* Effects of SMS and extracts of its component herbs on myocardial contractility in isolated normal rat hearts (30 min after drug administration). Herbal extracts: GS – *Radix Ginseng*; MD – *Radix Ophiopogonis*; FS – *Fructus Schisandrae*; SMS – Shengmai San.

*, ** $p < 0.05$ and $p < 0.01$, respectively, when compared with the control.

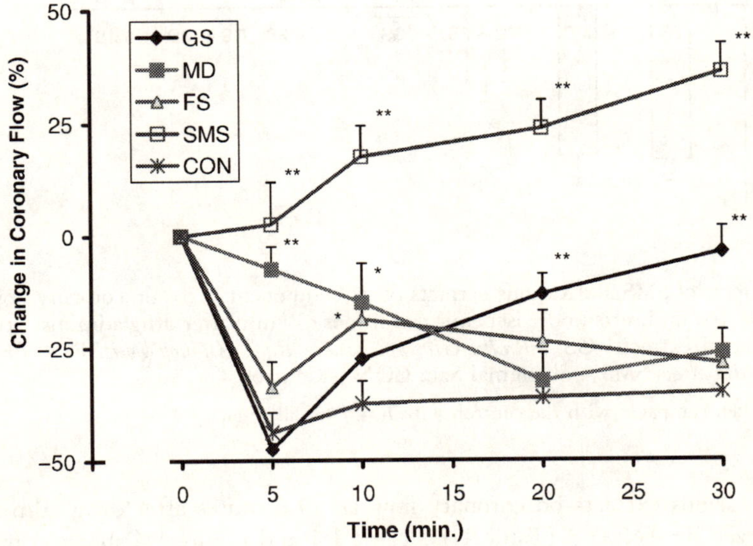


*Figure 1.3* Effects of SMS and extracts of its component herbs on coronary flow in isolated rat heart under ischemic condition. Herbal extracts: GS – *Radix Ginseng*; MD – *Radix Ophiopogonis*; FS – *Fructus Schisandrae*; SMS – Shengmai San; CON – Control.

*, ** $p < 0.05$ and $p < 0.01$, respectively, when compared with the control, n = 7 for all groups.

*Table 1.1* Effects of the three herbal components of SMS on coronary flow – ANOVA of a $2^3$ factorial design

| *Source of error* | *Degree of freedom* | *Sum of square* | *Mean square* | *F-value* | *P-value* |
|---|---|---|---|---|---|
| Total | 55 | 30726.9 | | | |
| Between group | 7 | 25775.7 | | | |
| GS | 1 | 17048.3 | 17048.3 | 165.3 | <0.01 |
| MD | 1 | 3478.6 | 3478.8 | 33.7 | <0.01 |
| FS | 1 | 2949.8 | 2949.8 | 28.6 | <0.01 |
| GS-FS | 1 | 280.4 | 280.4 | 2.72 | <0.05 |
| GS-MD | 1 | 283.1 | 283.1 | 2.74 | <0.05 |
| FS-MD | 1 | 1082.1 | 1082.1 | 10.49 | <0.01 |
| GS-MD-FS | 1 | 653.5 | 653.5 | 6.34 | <0.05 |
| Error | 48 | 4951.2 | 103.2 | | |
| $F_{0.05} = 4.04$ | | $F_{0.01} = 7.20$ | | | |

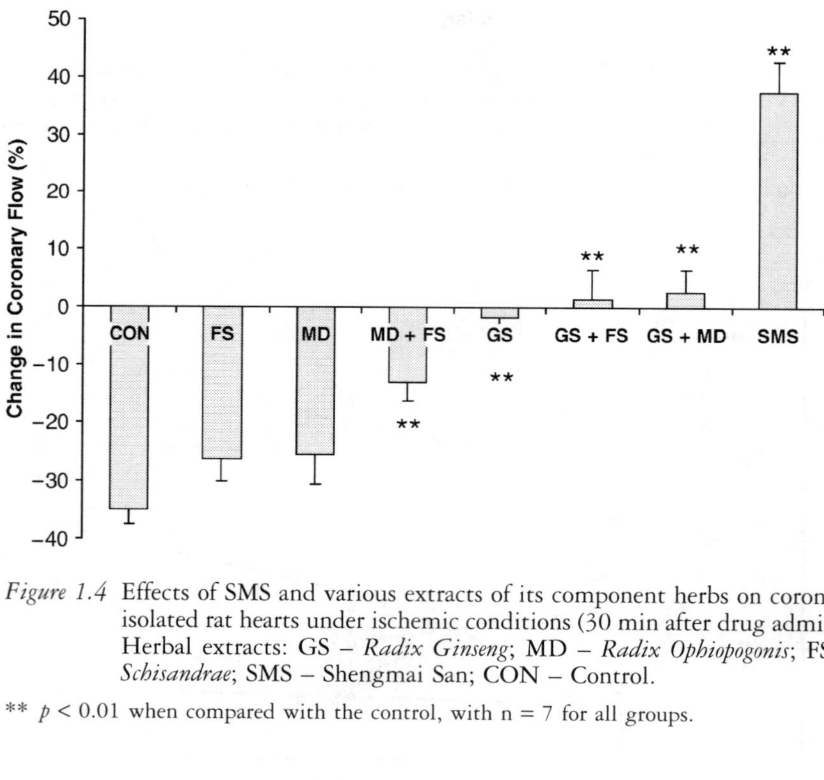


*Figure 1.4* Effects of SMS and various extracts of its component herbs on coronary flow of isolated rat hearts under ischemic conditions (30 min after drug administration). Herbal extracts: GS – *Radix Ginseng*; MD – *Radix Ophiopogonis*; FS – *Fructus Schisandrae*; SMS – Shengmai San; CON – Control.

** $p < 0.01$ when compared with the control, with n = 7 for all groups.

The effect of various extracts on coronary flow at 30 minutes after drug administration was analyzed by ANOVA (Table 1.1). Table 1.1 and Figure 1.4 shows that the three herbs have significant contribution ($p < 0.01$) to the pharmacological actions of SMS. Among them, *Radix Ginseng* had the most contribution, with F-value being greater than 165.3 and greater than those of other inter-group values, suggesting the leading role of *Ginseng* in the formulation. Interaction among the 3 individual herbs was

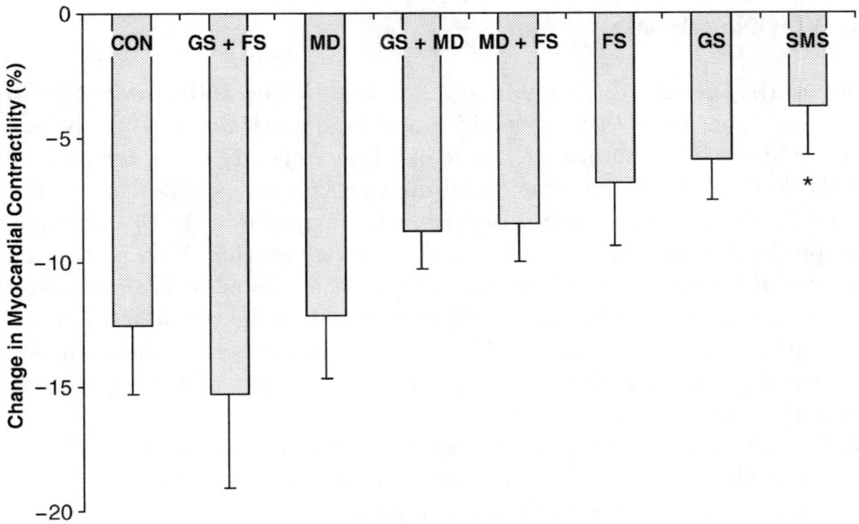


*Figure 1.5* Effects of SMS and various extracts of its component herbs on myocardial contractility of isolated rat hearts under ischemic conditions (30 min after drug administration). Herbal extracts: GS – *Radix Ginseng*; MD – *Radix Ophiopogonis*; FS – *Fructus Schisandrae*; SMS – Shengmai San; CON – Control.

** $p < 0.05$ when compared with the control, with n = 7 for all groups.

obvious ($p < 0.05$). The fact that the effect of SMS was stronger than that of the sum of individual herbs suggests a synergistic effect produced by the multi-component prescription.

### B. *Effect on myocardial contractility*

The effects produced by SMS and the three individual herbs on myocardial contractility were shown in Figure 1.5. At 30 minutes post-dosing, the action produced by SMS was most potent ($p < 0.05$).

In summary, the enhancing effect on coronary flow produced by SMS was stronger than the sum of those of the three individual herbs. SMS exhibited a sustained inhibitory action against coronary spasm induced by pituitrin. The significant interaction among the three components suggests a synergistic action produced by SMS, which provides a scientific basis for the multi-component formulation in TCM.

In TCM, the formulation of a prescription follows the rule of *Monarch*, *Minister*, *Assistant* and *Guide*. Experimental studies have demonstrated that the lowering of coronary resistance and enhancement of coronary flow by SMS is mainly afforded by *Radix Ginseng*, a *Monarch* herb in the prescription. While the effects produced by *Radix Ophiopogonis* (*Minister*) and *Fructus Schisandrae* (*Minister*) are not prominent, they can enhance the action of *Radix Ginseng*. These observations are in line with the theory in TCM that the *Minister* and *Assistant* serve to enhance the action of the *Monarch*.

## TRADITIONAL AND CONTEMPORARY CLINICAL APPLICATIONS OF SMS

According to the *Standard for Diagnosis and Treatment* (Wang 1602), SMS, comprising *Radix Ginseng* 25 g, *Radix Ophiopogonis* 15 g and *Fructus Schisandrae* 15 g, is decocted with water and can be consumed all year round. Contemporary use of the prescription include (1) decoction with water; and (2) injection (s.c. or i.v.), 2–4 ml each time, administered every 1–2 hours. SMS is prescribed for invigorating the *Qi* and promoting *body fluid* production, astringing the *Yin* and stopping sweating. SMS is indicated for the treatment of (1) deficiencies of the *Yin* and *Qi*, as manifested by lassitude, shortness of breath, reluctance of speech, thirst, hidrosis, dryness of the throat and tongue, and indistinct pulse; and (2) impairment of the *lung* due to persistent cough, impairment of the *Yin* and *Qi*, dry cough with little phlegm, shortness of breath, spontaneous sweating and asthenic pulse.

Apart from the treatment of diseases associated with deficiencies of the *Yin* and *Qi*, such as heat stroke, infantile summer fever, functional hypothermia and other acute feverish diseases, SMS has been frequently used in dealing with clinical conditions of heart failure, shock and cardiomyopathy. In addition, SMS, with minor alterations of the formula, may also be applied to tachycardia, angina, arrhythmia, neurasthenia, Keshan Disease, bronchitis associated with *Yin*- and *Qi*-deficiencies, and persistent cough associated with tuberculosis. If it is used for treating depletion of the *Yin*, the amount of *Radix Ginseng* in the prescription should be increased accordingly. On the other hand, as SMS possesses astringing action, it should not be used if the *exogenous evils* (pathogens) or *heat syndromes* (fever) prevail without accompanying impairment of the *Qi* and the *body fluid*.

## SMS-RELATED FORMULATIONS AND THEIR CLINICAL APPLICATION

### I. Modified Shengmai powder

Reference: *Analysis of Epidemic Febrile Diseases, Vol. 1* (Wu 1798)

Formulation: *Radix Adenophorae Stritae*, *Radix Ophiopogonis*, *Fructus Schisandrae*, *Cortex Mutan Radicis* and *Radix Remanniae*.

Functions: Cooling the *blood*, nourishing the *Yin*, promoting *body fluid* production.

Indications: latent summer-heat, warm mouth, hidrosis and red tongue.

### II. Shengmai drink

Reference: *Chinese Pharmacopoeia Vol. 1* (Ministry of Health, China 2000)

Formulation: *Radix Codonopsis Pilosulae*, *Radix Ophiopogonis* and *Fructus Schisandrae*

Functions: Invigorating the *Qi* and restoring pulse, nourishing the *Yin* and promoting the production of *body fluid* production.

Indications: Palpitation, shortness of breath, indistinct pulse, asthenic sweat.

## III. Ophiopogonis drink for clearing lung heat (*Mendong Qingfeiyin*)

Reference: *Differentiation on Endogenous and Exogenous Diseases* (Li 1213)

Formulation: *Radix Astragali*, *Radix Paeoniae Alba*, *Radix Glycyrrhizae*, *Radix Ginseng*, *Radix Ophiopogonis*, *Radix Angelicae Sinensis* and *Fructus Schisandrae*.

Functions: Supplementing the *Qi*, benefiting the *spleen*, clearing the *lung heat* and nourishing the *Yin*.

Indications: *Qi*-deficiency of the *spleen* and *lung*, *dry heat* due to *Yin*-depletion, asthmatic cough, non-traumatic hemorrhage, haematemesis, lassitude, weak and rapid pulse.

## IV. Cold decoction of *Radix Ophiopogonis* (*Maimendong Yinzi*)

Reference: *Secret Record of the Chamber of Orchids Vol. II* (Li 1336)

Formulation: *Radix Astragali*, *Radix Ophiopogonis*, *Radix Angelicae Sinensis*, *Radix Remanniae* Recens, *Radix Ginseng* and *Fructus Schisandrae*.

Functions: Invigorating the *Qi* and nourishing the *blood*.

Indications: Persistent hematemesis.

## V. Ginseng-Schisandrae decoction

Reference: *A Collection of Pediatrics Vol. III* (Chen 1750)

Formulation: *Radix Ginseng*, rinsed *Radix Paeoniae Alba*, *Poria*, *Fructus Schisandrae*, *Radix Ophiopogonis* from the Province of Zhang, *Radix Glycyrrhizae Praeparata, Rhizoma Zingiberis Recens* and *Fructus Ziziphi Jujubae.*

Functions: Invigorating the *Qi* and strengthening the *spleen*.

Indications: Persistent cough, deficiency of the *spleen*, visceroptosis, pale complexion and white lips.

## VI. Ginseng-Astragali powder

Reference: *Main Rules in Medical and Health Service Vol. V* (Luo, Yuan Dynasty)

Formulation: *Radix Ginseng*, *Radix Gentianae Macrophyllae*, *Poria*, *Rhizoma Anmarrhenae*, *Cortex Mori Radicis, Radix Platycodi, Radix Asteris, Radix Bupleuri, Radix Astragali, Cortex Lycii Radicis, Radix Remanniae Recens, Rhizoma Pinelliae, Radix Paeoniae Rubra, Radix Asparagi, Carapax Trionycis* and *Radix Glycyrrhizae Preparata*.

Functions: Invigorating the *Qi* and nourishing the *Yin*, suppressing cough and resolving phlegm.

Indications: *Yin* and *Qi*-deficiency; emaciation; lassitude; feverish sensation in the palms, soles and heart; hectic fever and spontaneous sweating; dryness of the mouth and throat; cough with bloody phlegm; chest and hypochondrium discomfort; red tongue with little fur; small and rapid pulse.

## VII. *Radix Astragali* powder

Reference: *Taiping Benevolent Prescriptions Vol. 70* (Wang *et al.* 992)

Formulation: *Radix Astragali, Cortex Lycii Radicis, Red Poria, Radix Ophiopogonis, Radix Paeoniae Rubra*, dried *Radix Remanniae Recens, Radix Ginseng, Radix Bupleuri, Radix Scutellariae, Radix Angelicae Sinensis, Radix Glycyrrhizae.*

Functions: Invigorating the *Qi* and nourishing the *blood*, nourishing the *Yin* and antipyretic.

Indications: Fever associated with consumptive diseases; emaciation; persistent pain in the limbs; lassitude; dryness of the mouth and tongue; anorexia; acratia; red tongue and small pulse.

## VIII. Ginseng-Astragali powder

Reference: *An Encyclopedia of Useful Prescriptions for Women* (Chen 1237)

Formulation: *Radix Ginseng, Radix Astragali (fried), Radix Puerariae, Radix Gentianae Macrophyllae, Red Poria, Radix Ophiopogonis, Rhizoma Anmarrhenae* and *Radix Glycyrrhizae.*

Functions: Invigorating the *Qi*, dipersing the *heat* and promoting the production of *body fluid.*

Indications: Restlessness and dryness of the mouth associated with fever.

## IX. Decoction for flaccid limbs (*Wuwei Tang*)

Reference: *A Summary on Medicine from Clinical Practice Vol. III* (Cheng 1732)

Formulation: *Radix Ginseng, Rhizoma Atractylodis Macrocephalae, Poria, Radix Glycyrrhizae, Radix Angelicae Sinensis, Semen Coicis, Radix Ophiopogonis, Cortex Phellodendri and Rhizoma Anmarrhenae.*

Functions: Supplementing the *spleen* and invigorating the *Qi*; nourishing the *Yin* and dispersing the *heat.*

Indications: Flaccid syndromes manifested as flaccid limbs and during early stages of diseases.

## X. Decocted paste of Polygonati essence (*Yuhua Jian*)

Reference: *Yi Chun Sheng Yi*

Formulation: *Rhizoma Polygonati Odorati, Fructus Schisandrae, Radix Ophiopogonis, Radix Adenophorae Strictae, Radix Codonopsis Pilosulae, Poria, Rhizoma Dioscoreae, Radix Dipsaci, Radix Achyranthis Bidentatae* and *Rhizoma Atractylodis Macrocephalae.*

Functions: Supplementing the *lung-Qi* and nourishing the *lung-Yin.*

Indications: *Yin-* and *Qi*-deficiency of the *lung*, walking incompetence.

## XI. Decoction of ten herbs for nourishing primordial Qi (*Shiquan Yuzhen Tang*)

Reference: *Records of Traditional Chinese and Western Medicine in Combination* (Zhang 1918–1934)

Formulation: Wild *Radix Codonopsis, Radix Astragali Recens, Rhizoma Dioscoreae Recens, Rhizoma Anmarrhenae, Radix Scrophulariae, fresh Os Draconis, fresh Concha Ostreae, Radix Salviae Miltiorrhizae, Rhizoma Sparganii* and *Rhizoma Zedoariae.*

Functions: Invigorating the *Qi* and nourishing the *Yin*, resolving *blood stasis* and astringing.

Indications: Exhaustion manifested as emaciation, squamous and dry skin, hiccough, asthma, dreaminess, spontaneous sweating, seminal emission; rapid, stringy, small and indistinct pulse.

## XII. Ginseng injection

Reference: *Jilin Province Pharmaceutical Standards, 1985*

Formulation: *Radix Ginseng* and glucose.

Functions: Invigorating the *Qi* to arrest depletion, promoting the production of *body fluid* and tranquilizing the mind.

Indications: Exhaustion, depletion, dizziness, palpitation and diabetes associated the insufficiency of the *primordial-Qi*.

Clinical Applications: Treatment of shock, neurasthenia, diabetes, cancers and hypotension.

## XIII. Ginseng-Ophiopogonis injection

Reference: *Valuable Prescriptions* (Sun, mid-7th century), Chui Provincial Pharmaceutical Approval Number 001552 (1981)

Formulation: *Radix Ginseng* and *Radix Ophiopogonis.*

Functions: Invigorating the *Qi* and promoting the production of *body fluid*, suppressing thirst and preventing exhaustion.

Indications: Deficiency of the *Qi* and depletion of *body fluid* induced by various diseases, manifested as symptoms of syncope and deficiencies, such as dizziness, spontaneous sweating, palpitation, thirst, indistinct pulse.

Clinical Applications: Shock, hypotension, arrhythmia, coronary heart disease, epidemic hemorrhagic fever and myocarditis.

## XIV. Astragali-Shengmai drink

Reference: *Differentiation on Endogenous and Exogenous Diseases*, Zhe Provincial Pharmaceutical Approval Number 050701 (1996)

Formulation: *Radix Astragali*, *Radix Codonopsis Pilosulae*, *Radix Ophiopogonis* and *Fructus Schisandrae*.

Functions: Invigorating the *Qi* and tonifying the *Yin*, nourishing the *heart* and supplementing the *lung*.

Indications: Deficiency of the *Yin* and *Qi*, palpitation and shortness of breath associated with coronary heart disease and senility.

Clinical Applications: Coronary heart disease and debility associated with old age.

## XV. Treasure for preserving youth (*Qingchun Bao*)

Reference: *Zhejiang Province Pharmaceutical Standards (1985)*

Formulation: *Radix Ginseng, Radix Asparagi, Radix Rehmanniae*, etc.

Functions: Invigorating the *Qi* and nourishing the *blood*, tonifying the *Yin* and promoting the production of *body fluid*, tranquilizing the mind and anti-aging.

Indications: Deficiency of the *Qi* and hemophthisis, lassitude, asthma associated with physical exercise, anorexia; palpitation, amnesia, insomnia; general debility due to sickness or old age.

Clinical Applications: Middle-aged people who consume the prescription regularly could see improvement in energy, mentality, endurance and adaptability to environment. It also supports the immune system and improves the body's resistance to diseases.

## XVI. Hypoglycemic agent (*Taizi Bao*) (experimental formula)

Formulation: *Fructus Schisandrae*, *Radix Berberidis*, *Cuculus Poliocephalus* from *Xingan*, *Cortex Syringae*, *Radix Stemonae*, *Rhizoma Zingiberis* and *Radix Ginseng*.

Functions: Tonifying and nourishing the *five viscerae*, promoting *body fluid* production and quenching thirst, suppressing abnormalities in *stomach* and its regulation.

Indications: Diabetes.

## XVII. Tonic for benefiting the heart (*Zhenqi Fuzheng Tang*) (empirical formula)

Reference: Pharmaceutical Approval Number Z-64 (1992)

Formulation: *Radix Ginseng*, *Radix Ophiopogonis*, *Fructus Schisandrae* and *Rhizoma Anmarrhenae*, etc.

Functions: Invigorating the *heart-Qi*, nourishing the *heart-Yin* and improving circulation.

Indications: Deficiency of the *heart-Qi* or deficiencies of the *Yin* and *Qi*; chest pain associated with atherosclerosis.

Clinical Applications: Coronary heart disease and angina.

**REFERENCES**

Anonymous (1186) *Origin of Medicine.*
Anonymous (Warring States) *Canon of Internal Medicine.*
Chen, F.Z. (1750) *A Collection of Pediatrics.*
Chen, Z.M. (1237) *An Encyclopedia of Useful Prescriptions for Women.*
Cheng, G.P. (1732) *A Summary on Medicine from Clinical Practice.*
Li, G. (1213 or 1247) *Differentiation on Endogenous and Exogenous Diseases.*
Li, G. (~1336) *Secret Record of the Chamber of Orchids.*
Luo, T.Y. (Yuan Dynasty) *Main Rules in Medical and Health Service.*
Ministry of Health, China (2000) *Chinese Pharmacopoeia.*
Sun, S.M. (mid-7th century) *Valuable Prescriptions.*
Wang, H.Y. *et al.* (992) *Taiping Benevolent Prescriptions.*
Wang, K.T. (1602) *Standard for Diagnosis and Treatment.*
Wu, J.T. (1798) *Analysis of Epidemic Febrile Diseases.*
Zhang, X.C. (1918–1934) *Records of Traditional Chinese and Western Medicine in Combination.*
Zhu, D.N., Li, Z.M., Yan, Y.Q., and Zhu, J.G. (1998a) Studies on the relationship between the dynamic changes in chemical composition and the pharmacological activities of Shengmai San: chemistry of SMS II, *China Journal of Chinese Materia Medica*, 23, 291–3.
Zhu, D.N., Li, Z.M., Yan, Y.Q., and Zhu, J.G. (1998b) Studies on the relationship between the dynamic changes in chemical composition and the pharmacological activities of Shengmai San 6: changes in the composition of ginsenoside, *Academic Periodic Abstracts of China*, 4, 869–71.

# 2 Pharmacological studies on Shengmai San

*Kam-Ming Ko, Duncan H.F. Mak, Tze-Kin Yim and Yong-Qing Yan*

## INTRODUCTION

Shengmai San (SMS) is a well-known traditional Chinese medicine formula, and has been used in China for more than eight hundred years. SMS is comprised of *Radix Ginseng* (*Panax ginseng C.A. Meyer*), *Radix Ophiopogonis* (*Ophiopogon japonicus Ker-Gawl*) and *Fructus Schisandrae* (*Schisandra chinensis (Turcz.) Baill*) and is prescribed for replenishing the *Qi* (*vital energy*) and promoting the production of *body fluid*, retaining the *Yin* and halting the excessive perspiration. Traditionally, SMS is used for the treatment of body disorders including: (1) heat-induced damage of the *Qi* and the deficiency of both the *Qi* and the *Yin* caused by the depletion of the *Yin-fluid*, with symptoms of profuse sweating, thirst, throat dryness, dyspnea, tiredness, weak pulse, red and dry tongue without saliva, and (2) *heart* and *lung* weakness and the deficiency of the *Qi* and the *Yin* due to protracted illness, with symptoms of cough with little phlegm, shortness of breath with spontaneous perspiration, dry mouth and hot tongue, and weak and faint pulse. Concomitant with a huge amount of clinical research having been done since the 1970s, SMS is commonly used in clinical practices with a prudent therapeutic efficacy, and its area of clinical application is being rapidly extended. Nowadays, SMS is clinically used for the treatment of shock caused by cardiovascular diseases, contractile heart failure, and myocardial ischemia etc. With regard to pharmacological studies, a lot of experimental work has also been done on SMS. In one of the large-scale cooperative research projects, SMS (prepared by standardized manufacturing procedures) was used for investigating the effect on cardiovascular system. Four main categories of experimental models were adopted, including those for contractile heart failure, myocardial ischemia, arrhythmia and shock. Experimental methods of different levels and standards were used for investigations, in which various physiological and pharmacological markers were measured. Other pharmacological studies on SMS have demonstrated the effect on experimentally-induced atherosclerosis and blood lipid content, the stimulatory action on immune system, the prevention against anaphylaxis, tissue injury and mutation induced by toxins, and the effect on the central nervous system. Recently, antioxidant activities of SMS have been investigated, which further provide a pharmacological rationale for preventive and therapeutic effect of SMS on cardiovascular diseases as well as for health safeguarding.

## CARDIOVASCULAR SYSTEM

## I. Myocardial ischemia

### *A. Effect on pituitrin-induced changes in coronary flow in isolated rat hearts*

Isolated rat heart was perfused with pituitrin-containing perfusate (40 u/L). After 2-min of perfusion, the coronary flow was significantly reduced, with the lowest coronary flow being achieved after 5 min of perfusion. The coronary flow was further reduced by 35 per cent after 30 min of perfusion. The supplementation of SMS in the perfusate (0.5 g/ml) could significantly enhance the coronary flow by 71.8 per cent, indicating the ability of SMS to antagonize the coronary spasm caused by pituitrin (Fig. 2.1) (Wang and Li 1990).

### *B. Effect on isoproterenol-induced myocardial damage in rats*

Rats were treated orally or injected intraperitoneally with SMS at a daily dose of 11/12 or 2.75/5.5 g/kg, respectively, consecutively for 8 or 10 days. One hour after the last dose, the animals were injected intraperitoneally with urethane (0.4 g/kg) for basic anesthesia, and then completely anesthetized with diethyl ether. ECG of channel II and five parallel channels attached to the thoracic region were recorded for comparison. Isoproterenol was given to animals by both intraperitoneal and subcutaneous injection at a dose of 2 mg/kg. ECG was continuously recorded for 30 min after the injection. After 24 hours the second injection of isoproterenol was given at the same dosage, and the ECG prior to and after the drug injection was also recorded. The animals were then sacrificed and blood samples were drawn for measuring plasma creatine phosphokinase

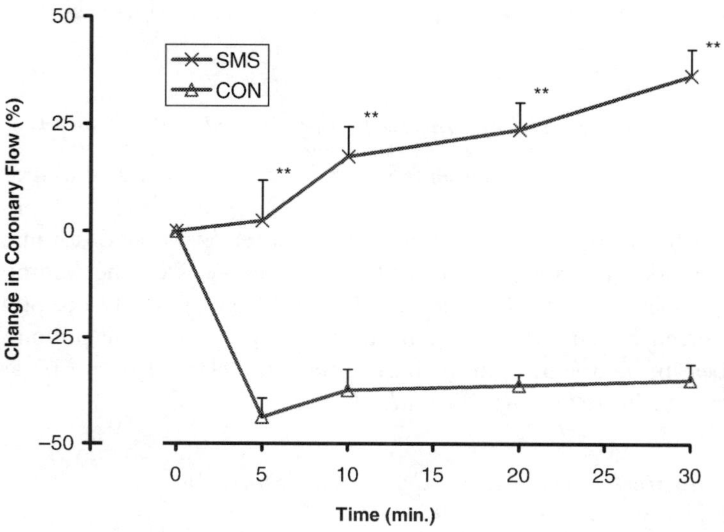


*Figure 2.1* Effects of SMS on coronary flow in isolated rat heart under ischemic condition. Herbal extracts: SMS – Shengmai San; CON – Control.

* $p < 0.05$, ** $p < 0.01$ when compared with the control, with n = 7 for all groups.

(CPK) activity. Injection of rats with isoproterenol could cause myocardial damage, as indicated by the change in ST segment of the ECG and the elevated plasma CPK activity. Oral pretreatment with SMS at the indicated dose could significantly lower the elevation and reduce the total shift of the ST segment, while intraperitoneal injection of SMS could significantly reduce the total shift of the ST segment as well as the plasma CPK activity (Zhang *et al.* 1990).

## II. Contractile heart failure

### *A. Effect on pentobarbital-induced contractile heart failure in dogs*

In a canine model of acute contractile heart failure (Li *et al.* 1990a), administration of pentobarbital (30 mg/kg, iv) could significantly change most of the hemodynamic parameters. While LVSP, LVdp/$dt_{max}$ and MAP were reduced by 70, 80 and 70 per cent, respectively, R-LVdp/$dt_{max}$ was increased by 12 per cent. Heart rate and cardiac output were reduced by 20 and 50 per cent, respectively. After the induction of acute contractile heart failure, intravenous infusion of SMS (0.5–2.0 g/kg; 0.08 ml/min/kg) for 60 min could increase LVSP, LVdp/$dt_{max}$, MAP, heart-rate and cardiac output, but did not produce any effect on R-LVdp/$dt_{max}$. When compared with SMS, ouabain (60 μg/kg) showed a more apparent and efficacious positive inotropic effect. Recent research indicated that the action mechanism of SMS in producing the myocardial effect involved a multiplicity of events. These include the suppression of cardiomyocyte membrane $Na^+/K^+$-ATPase activity (Beijing College of Traditional Chinese Medicine 1973), the enhancement on cellular anabolism for protection against ischemic and anoxic myocardial damage and the elevation of tissue glycogen and ATP content, the enhancement on DNA and RNA synthesis, and the maintenance of cellular membrane integrity under conditions of toxin-induced myocardial damage (Liu *et al.* 1978). The ability of SMS to produce beneficial modulatory effect on myocardial metabolism under the condition of myocardial injury was superior to the action produced by ouabain (Li *et al.* 1990a).

### *B. Effect on pentobarbital-induced contractile failure in isolated rat hearts*

Isolated heart was given perfusate supplemented with pentobarbital at a concentration of 0.5 g/L to induce failure and cardiac arrest. When being perfused back with physiological perfusate, the heart could restore beating. SMS, when given at doses ranging from 0.5–1.0 g/L, could delay the onset of cardiac arrest and hasten the resumption of beating following cardiac arrest, while ouabain (0.1–1.0 μmol/L) could not produce any effects. When given at same dosage again, SMS could significantly prolong the duration of heart beating and significantly shorten the time elapsed between cardiac arrest and resumption of heart beating (Li *et al.* 1990c).

### *C. Effect on pentobarbital-induced acute contractile failure in cultured rat cardiomyocytes in vitro*

The myocardial effect of SMS was examined using a cellular model of pentobarbital-induced acute contractile failure. Ouabain (0.17 μmol/L) could significantly counteract the pentobarbital-induced contractile failure in cultured cardiomyocytes, but the protection was associated with arrhythmic contraction. SMS, when added at a final concentration of 1 mg/ml, could restore the contractile function of cardiomyocytes, with a normal frequency and rhythm of beating. Both ouabain and SMS treatment could significantly delay the onset of cardiac arrest induced by pentobarbital (Li *et al.* 1990b).

## III. Cardiac arrhythmia

### *A. Effect on cardiac arrhythmia induced by electrical stimulation on rabbit hypothalamus*

Electrical stimulation on rabbit hypothalamus caused frequent supraventricular and ventricular arrhythmic response. Oral treatment of SMS at a dose of 4.4 g/kg could significantly prevent the cardiac arrhythmia caused by electrical stimulation on the hypothalamus (Hou *et al.* 1990).

### *B. Effect on ventricular fibrillation induced by chloroform in mice*

After being anesthetized with chloroform, the rate of endoventricular fibrillation in mice was observed for 30s. Intraperitoneal injection of SMS at doses ranging from 8.25–13.75 g/kg could significantly protect against the ventricular fibrillation induced by chloroform (Hou *et al.* 1990).

### *C. Effect on cardiac arrhythmia induced by calcium chloride in rats*

Rapid intravenous perfusion of 10 per cent calcium chloride solution at a dose of 0.15 g/kg caused lethal ventricular fibrillation in rats. When given SMS by intraperitoneal injection at a dose of 2.75 g/kg, the rats were protected against the lethal effect of calcium chloride, as indicated by the reduction of mortality rate from 100 to 20 per cent (Hou *et al.* 1990).

### *D. Effect on cardiac arrhythmia induced by electrical stimulus to isolated guinea-pig hearts*

Electrical stimulation (16HZ; 7–14V) could induce ventricular fibrillation in isolated and perfused guinea-pig hearts. Perfusate supplemented with SMS (0.11 g/min) could produce an anti-fibrillation rate of 66.7 per cent (Hou *et al.* 1990).

## IV. Shock

### *A. Hemorrhagic shock*

#### *(1) Effect on hemorrhagic shock in rabbits*

A rabbit model of hemorrhagic shock was established by blood letting till the blood pressure was dropped to a level under 5.3 kPa (40 mmHg) and the lowered blood pressure could sustain under 8 kPa (60 mmHg) for at least 20 min. Within 1 to 5 min after the intravenous perfusion with SMS at a dose of 2.2 g/kg, the blood pressure started to elevate. The duration of elevated blood pressure was maintained for 93.0 ± 16.4 min. When compared with the control, SMS could significantly protect against the hemorrhagic shock, with the period of stable blood pressure being longer than that of dextran (Chu *et al.* 1990).

#### *(2) Effect on hemorrhagic shock in dogs*

Intravenous injection with SMS at a dose of 2.2 g/kg could produce a significant elevation in blood pressure at the early stage of acute hemorrhagic shock in the canine model (Chu *et al.* 1990).

#### *(3) Effect on ischemia-induced acute cardiac arrest in rats*

The time taken to reach the onset of ischemia-induced cardiac arrest was observed in blood-letting rats. SMS, given at a dose of 5.5 g/kg by either intraperitoneal or oral administration could significantly delay the onset of cardiac arrest in ischemic hearts, indicating the ability of SMS to protect against the hemorrhagic shock.

### *B. Endotoxin shock*

#### *(1) Effect on endotoxin-induced shock in rabbits*

After the intravenous injection with purified E coli-derived endotoxin at a dose equivalent to 1.5 times of $LD_{50}$, the rabbit model of endotoxin shock was successfully established when the blood pressure was dropped to less than one-third of the normal level within 90 min. When given at a dose of 2.2 g/kg by intravenous injection, SMS could produce significant protective effect on the endotoxin-induced shock, but failed to reduce the mortality rate (Chu *et al.* 1990).

#### *(2) Effect on endotoxin-induced shock in mice*

Mice were pretreated orally with SMS at a daily dose of 5.5 g/kg for 3 days. One hour after the last dose, the animals were given E. coli-derived endotoxin by intravenous injection at a dose equivalent to 50 per cent of $LD_{50}$. The same dosage of endotoxin was given again 6 hr later. The mortality rate of animals in 24 hr was observed. Results from this study indicated that SMS could significantly reduce the mortality rate caused by the endotoxin shock in mice (Chu *et al.* 1990).

### *C. Cardiogenic shock*

#### *(1) Effect on acute heart dysfunction in rabbits*

An animal model of acute heart dysfunction was established by intravenous injection of olive oil into rabbits (Zhao *et al.* 1974). SMS was given by intravenous injection at a dose of 2.2 g/kg 30 min prior to the injection of olive oil. Results from this study indicated that SMS could significantly protect against the acute heart dysfunction induced by olive oil-mediated lung occlusion, which was suggestive of preventive action in cardiogenic shock (Chu *et al.* 1990).

*(2) Effect on experimentally-induced cardiogenic shock in dogs*

An open-chested canine model of cardiogenic shock was prepared by the ligation of the anterior descending branch of coronary artery (Beijing Medical University 1974, 1975). The animal model was successfully established when the blood pressure was decreased to a level below 9.33 kPa (70 mmHg) after the coronary ligation, with this condition being sustained for 15 min. Intravenous dripping for 1 hr or intragastric administration of SMS at a dose of 5.5 g/kg caused the recovery of the blood pressure to a level exceeding that at the early phase of the shock, which was sustained above 9.33 kPa (70 mmHg) for 1 or 1.5 hr, respectively, following the drug administration. This result indicated the significant therapeutic effect produced by SMS on cardiogenic shock (Chu *et al.* 1990).

## V. Atherosclerosis

Recently, it has been reported that inhibitors of blood platelet aggregation could inhibit the formation of experimentally-induced atherosclerotic plaque (Smith and Hilker 1990). Given the ability of SMS to suppress the aggregation and release of blood platelets, the effect of SMS on atherosclerosis is described below.

### *A. Effect on experimentally-induced atherosclerosis*

A rabbit model of atherosclerosis was prepared by feeding the animals daily with diet supplemented with 1 g of cholesterol, consecutively for 6 weeks. Results obtained from atherosclerotic (As) group, SMS-treated-As group and paeonol-treated-As group (positive control) were compared with the control group (i.e. no cholesterol supplementation). In the drug treatment group, animals were given SMS intragastrically or paeonol intraperitoneally at a daily dose of 2.5 g/kg or 100 mg/kg respectively, for 6 days per week and consecutively for 6 weeks. Microscopic examination on pathological changes and the area of damage in aortic intima as well as histochemical analysis on pathological tissues. This indicated that both SMS and paeonol treatment could ameliorate the severity of pathological changes in the aorta, with reduced extents of intimal proliferation, lipid disposition, as well as lower lipase activity and extent of mucopolysaccharide formation, when compared with the control group. These observations suggested that SMS could significantly inhibit the formation of As plaques in aortic intima. Some aortic smooth muscle cells in As animals was found with their nuclei perpendicular to the intima, and the smooth muscle cells of the middle layer were found to invade the endothelial layer at injured site of the elastic membrane. In contrast, the nuclei of smooth muscle cells in SMS-treated animals were parallel with the elastic membrane. These observations suggested that SMS could inhibit the movement of arterial smooth muscle cells from the middle layer to the endothelial layer, which may be related to the inhibitory action on As plaque formation. The ensemble of results indicated that SMS could suppress the As-induced pathological changes in the aorta, including intimal proliferation, connective tissue formation and lipid deposition, to varying extents (Shi *et al.* 1990).

*Table 2.1* Effect of SMS on serum lipids and lipoprotein cholesterol levels

| | *Blood lipid content* (mg/100ml, mean ± S.D.) | | | | | | *HDLc/TCh* | *AsI* |
|---|---|---|---|---|---|---|---|---|
| | *TCh* | *TG* | *LDLc* | *HDLc* | *HDL2c* | *HDL3c* | *(ratio)* | |
| CON | 757.9 ±29.9 | 105.0 ±45.8 | 702.6 ±34.5 | 2.74 ±7.0 | 11.0 ±1.3 | 16.4 ±6.3 | 0.036 ±0.009 | 0.964 ±0.009 |
| SMS | 557.7** ±137.6 | 85.4** ±42.0 | 500.8** ±127.3 | 39.8** ±6.3 | 16.7** ±4.3 | 23.1** ±3.2 | 0.073** ±0.013 | 0.925** ±0.010 |
| Paeonol | 749.2 ±35.2 | 161.2 ±67.5 | 683.2 ±21.0 | 29.8 ±6.0 | 10.5 ±4.0 | 19.3 ±4.9 | 0.040 ±0.010 | 0.960 ±0.007 |

** $p < 0.01$ when compared with the control (CON)

## *B. Effect on the levels of serum lipids and cholesterol in lipoproteins in a rabbit model of atherosclerosis*

### *(1) Levels of serum lipids and cholesterol in lipoproteins*

A rabbit As model was prepared by feeding cholesterol in the diet for 6 weeks. Venous blood samples was taken from the ear ring region prior to the beginning of the first week, on the fourth week and sixth week of cholesterol feeding. Blood samples were measured for serum levels of total cholesterol (TCh), triglycerides (TG), low-density lipoprotein cholesterol ($LDL_C$), high-density lipoprotein cholesterol ($HDL_C$) and its subgroup cholesterol ($HDL_{2C}$, $HDL_{3C}$). On the fourth and sixth week after the cholesterol feeding, serum TCh and $LDL_C$ levels were significantly elevated, while $HDL_C$ and $HDL_{2C}$ levels were significantly lowered when compared with the respective level prior to cholesterol feeding. All these indicated the successful induction of hyperlipidemic status. SMS treatment at a daily intragastric dose of 2.5 g/kg for 6 days per week, consecutively for 6 weeks, could significantly lower serum TCh, and TG, as well as $LDL_C$ levels, but increase the $HDL_C$ and $HDL_{2C}$ levels, when compared with the As control group. Paeonol, when given by intraperitoneal injection at a daily dose of 100 mg/kg for 6 weeks, did not produce any detectable change in the levels of serum lipids and lipoprotein cholesterol (Table 2.1) (Shi *et al.* 1990).

### *(2) As inducing index (AsI) and Anti-As index ($HDL_c$/TCh)*

It is widely accepted that both $HDL_c$ and $HDL_{2c}$ levels are the anti-As factors while $LDL_c$ is regarded as As inducing factor. However, recent studies have demonstrated that the As inducing index (AsI, TCh-$HDL_C$)/$HDL_C$) and Anti-As index ($HDL_c$/TCh) would be a more sensitive parameter for the evaluation or forecast of incidence of As-related cardiovascular diseases. Effect of SMS on these parameters in the As rabbit model was examined. The results indicated that SMS treatment (2.5 g/kg, 6 days/week, 6 weeks) could significantly lower the AsI and elevate the Anti-As index (Table 2.1) in the As rabbit model, when compared with the control.

According to the present experimental results, the mechanism involved in the suppression of As plaque formation by SMS may be related to the following actions: (i) inhibition of blood platelet aggregation which, in turn, suppresses the release of active substances like platelet-derived growth factor, (ii) the modulation on the metabolism of blood lipids, (iii) the suppression of As-inducing factor like $LDL_C$ level and (iv) the enhancement of anti-As factors like $HDL_c$ and $HDL_{2c}$ levels (Shi *et al.* 1990).

## VI. Toxin-induced myocardial injury

### *A. Effect on mitomycin (MMC)-induced myocardial ultrastructural changes in mice*

From three days onwards after the intraperitoneal injection with MMC at a dose of 5 mg/kg, the mice showed an abrupt decrease in body weight, reduction in physical activities, messy body hairs, decrease in food intake and deep yellowish turbid urine. SMS treatment at a daily dose of 10 g/kg for 3 days could slightly ameliorate the progressive decrease in body weight of MMC-treated animals. Within 10-day period of observation, the mortality rate in MMC-treated animals was 20 per cent, while all SMS-treated animals survived after the MMC challenge. Electron microscopic examination showed that SMS could significantly protect against the MMC-induced damage on the nucleus and myofibril of cardiomyocytes. In addition, SMS treatment also produced a certain degree of protection to mitochondria and other cellular organelles (Sun *et al.* 1990).

### *B. Effect on MMC-induced morphological and functional changes in cultured rat cardiomyocytes*

MMC did not produce any immediate effect on the beating performance of cardiomyocytes, however, the degree of influence was increased with time. Six hours after the addition of MMC at a final concentration of 1 mg/ml, the beating performance of cardiomyocytes was attenuated, with the reduced amplitude of contraction being associated with a lowered frequency, but normal rhythm of beating. Twenty-four hours after the administration of MMC, the beating performance was further attenuated, with a further decrease in beating frequency associated with abnormal rhythm. A small number of cardiomyocytes stopped beating, with morphological changes such as the decrease in pseudocytoplasmic extensions that had wrinkled and shrunk into a circular shape. Forty-eight hours after the MMC administration, most of the cardiomyocytes had stopped beating or even detached and died. While SMS treatment did not produce any detectable effect on the contractile function of normal cardiomyocytes, it significantly protected against the MMC-induced deterioration on beating performance. When added at a final concentration of 125 mg/ml, SMS increased the beating frequency by 42.2 per cent, when compared with the MMC-challenged control (Sun *et al.* 1990).

### *C. Effect on MMC-induced lactate dehydrogenase (LDH) leakage from cultured cardiomyocytes in vitro*

LDH is an intracellular enzyme present abundantly in myocardial cells. Under normal conditions, cultured cardiomyocytes only release trace amounts of LDH into the culture medium. However, when the cells are injured, the increase in membrane permeability will cause a significant LDH leakage (Higgins *et al.* 1980). Experimental results demonstrated that the extent of LDH leakage, indicative of cellular damage, was exaggerated

*Table 2.2* Protective effect of SMS on MMC-induced myocardial injury

| | *Dosage (μg/ml)* | *Number of samples* | *LDH (U, mean ± S.E.)* |
|---|---|---|---|
| Control | | 6 | 32.8 ± 6.9 |
| MMC | 5 | 6 | 57.5 ± 3.2 |
| MMC + SMS | 250 | 6 | 65.0 ± 8.3 |
| MMC + SMS | 500 | 6 | 58.5 ± 4.9 |
| MMC + SMS | 100 | 6 | 38.1 ± 7.8* |

* $p < 0.05$ when compared with the MMC group

*Table 2.3* Effects of SMS on MMC-induced inhibition of DNA synthesis in cardiomyocytes

| *Dosage (mg/ml)* | *No. of cell bottle* | *CPM (mean ± S.E.)* | *DNA synthesis (% change)* |
|---|---|---|---|
| Control | 7 | 2588.1 ± 94.4 | |
| MMC (1) | 7 | 1290.6 ± 67.6 | 50.5 |
| MMC + SMS (1) | 7 | 1218.0 ± 130.7 | 47.6 |
| MMC + SMS (125) | 7 | 1443.3 ± 127.8 | 56.4 |
| MMC + SMS (250) | 7 | 1562.6 ± 92.0* | 61.8 |

* $p < 0.05$ when compared with the MMC group

by increasing doses of MMC. SMS, when given at a concentration of 1 mg/ml, could significantly reduce the LDH leakage from MMC-challenged (5 mg/ml) cardiomyocytes, whereas lower doses (250 or 500 mg/ml) of SMS did not produce any significant effect (Table 2.2) (Sun *et al.* 1990).

### *D. Effect on MMC-induced changes in DNA synthesis cultured rat cardiomyocytes in vitro*

Cardiomyocytes were isolated from neonatal rats for culture. On the third or fourth day of culture, active cellular division and rapid DNA synthesis accompanied by maximum $^3$H-TdR cellular uptake were observed (Blodel *et al.* 1971). Experimental results demonstrated that the administration of MMC could decrease the $^3$H-TdR cellular uptake in a dose-dependent manner, with more than 90 per cent of inhibition at doses greater than 1 mg/ml. While SMS treatment only slightly enhanced the DNA synthesis in the purified cardiomyocytes culture, it could produce significant protective effect on MMC-induced inhibition on DNA synthesis. When added at final concentrations of 125 and 250 mg/ml, SMS could significantly increase the cellular uptake of $^3$H-TdR, with the effect being more apparent at the high dose, when compared with the MCC-challenged control (Table 2.3) (Sun *et al.* 1990).

### *E. Effect on adriamycin (ADR)-induced myocardial ultrastructural changes in rats*

Rats were administered intragastrically with ADR at a weekly dose of 2 mg/kg, consecutively for 10 weeks, pathological ultrastructural changes were found in myocardial cells isolated from the animals. SMS treatment at an oral dose of 3 g/day for 10 weeks could prevent the ADR-induced pathological ultrastructural changes in the rat heart (Rong *et al.* 1983).

### *F. Effect on ADR-induced changes in cell division and protein synthesis in cultured cardiomyocytes in vitro*

Experimental results indicated that SMS, when added at a final concentration of 1 mg/ml, increased the cell division index of cardiomyocytes after the ADR-challenge. SMS also enhanced the rate of protein synthesis in ADR-treated cardiomyocytes (Zhang 1996).

### *G. Effect on ADR-induced $Ca^{2+}$ overload in cultured cardiomyocytes in vitro*

Using fluorescent method for the detection of intracellular $Ca^{2+}$ level, it was found that ADR could obviously increase the intracellular $Ca^{2+}$ level in cardiomyocytes. SMS treatment could effectively suppress the ADR-induced overloading of intracellular $Ca^{2+}$ (Zhang 1996). Experimental results also indicated that SMS treatment could inhibit the alterations in sacroplasmic reticulum-mediated absorption and release of $Ca^{2+}$ caused by ADR (Zhang 1996).

## VII. Viral myocarditis

Coxsackievirus $B_3$ was used to reproduce a Balb/c mouse model of viral myocarditis. The intraperitoneal injection of the virus to the animals was accompanied by a simultaneous administration of SMS at a daily dose of 11 g/kg, consecutively for 10 days. SMS treatment significantly increased the survival rate of the viral intoxicated mice, with a decrease in serum LDH activity, but no detectable changes in serum CPK and aspartate aminotransferase (AST) activities. SMS treatment also produced a slight but not significant protective effect on the pathological changes (i.e. necrosis, calcification and infiltration) in myocardial tissue obtained from the infected animals (Table 2.4) (Zhuang *et al.* in press).

*Table 2.4* Effects of SMS on serum enzyme activities and survival rate in $CVB_3$-infected mice at day 10 of injection

| | *CPK (U/L)* | *LDH (U/L)* | *AST (U/L)* | *Survival rate (%)* |
|---|---|---|---|---|
| Non-infected control | 315.6 ± 112.2 | 2334.4 ± 355.4 | 42.9 ± 15.8 | 100 |
| $CVB_3$ control | 555.4 ± 192.4 | 3376.4 ± 457.5‡ | 154.6 ± 31.0‡ | 64† |
| $CVB_3$-SMS (11 g/kg) | 504.4 ± 184.3 | 2857.8 ± 494.4* | 233.6 ± 81.4 | 100* |

Each value is the mean ± S.D. with n = 7–10

* $p < 0.05$ when compared with the $CVB_3$ control

†, ‡ $p < 0.05$ and $p < 0.01$ respectively, when compared with the non-infected control

*Table 2.5* Inhibitory effects of SMS on voluntary mobility of mice

| | *Dosage* | *Number of experimental group (number of mice)* | *Mobility number (mean ± S.D.)* | *Inhibitory rate (%)* |
|---|---|---|---|---|
| Control | 20 ml/kg | 14 (28) | 945.3 ± 181.0 | |
| SMS | 5.5 g/kg | 14 (28) | 885.7 ± 178.2 | 6.3 |
| | 11.0 g/kg | 14 (28) | 698.8 ± 130.6* | 26.1 |
| | 22.0 g/kg | 14 (28) | 598.8 ± 133.3* | 36.6 |
| Chlorpromazine | 2.5 mg/kg | 14 (28) | 565.8 ± 162.3* | 40.1 |

* $p < 0.01$ when compared with the control

## CENTRAL NERVOUS SYSTEM (CNS)

### I. Mice behavior activities

#### *A. Effect on voluntary mobility in mice*

Using a three-light path mice activities detecting machine (Yuan *et al.* 1985), the activity index of mice in various groups was simultaneously detected, and each experiment was repeated 14 times. Experimental results indicated that intragastric administration of SMS at a dose of 11 or 22 g/kg could inhibit the voluntary mobility by 26.1 or 36.6 per cent, respectively. Under the same experimental conditions, chlorpromazine (2.5 mg/kg, ig) inhibited the voluntary mobility by 40 per cent. These indicated that SMS could produce an inhibitory effect on the voluntary mobility in mice (Table 2.5) (Yuan and Zhang 1990).

#### *B. Effect on passive activities in mice*

The experiment was performed using a bar rotation method. Mice were trained to run on a rotary bar rotating at fifteen revolutions per min. Mice capable of running on the rotary bar for more than 10 min without falling down were chosen for the experiment. Fifty minutes after the intragastric administration of SMS at doses ranging from 11–33 g/kg, the number of mice falling down was recorded. Results from this study indicated that SMS treatment did not produce a significant effect on muscle strength or motor coordination. The ability of SMS to inhibit voluntary mobility, as mentioned in previous section, was unlikely due to its effect on affecting the motor ability of the animals. Similarly, chlorpromazine (4 mg/kg, ig) did not produce any detectable effect on passive activities in mice (Yuan and Zhang 1990).

#### *C. Effect on fighting and aggressive behaviors in mice*

Male mice were maintained alone in a pottery can for more than one week. Animals showing fighting and aggressive behaviors were chosen for the experiment. Experimental results indicated that intragastric administration of SMS at a dose of 2.75 g/kg could produce a suppressive effect on the fighting and aggressive behaviors in mice, but the degree of suppression was smaller than that produced by chlorpromazine (5.0 mg/kg) (Table 2.6).

*Table 2.6* Inhibitory effects of SMS on aggressive behavior in mice

| | *Dosage* | *No. of mice* | *Disappearance of aggressive behavior (No.)* |
|---|---|---|---|
| Control | 25 ml/kg | 10 | 0 |
| SMS | 13.75 g/kg | 10 | 1 |
| | 27.5 g/kg | 10 | 6* |
| Chlorpromazine | 5.0 mg/kg | 10 | 10** |

*, ** $p < 0.05$ and $p < 0.01$ respectively, when compared with the control

*Table 2.7* Synergistic effect of SMS on the sedative-hypnotic action of pentobarbital

| | *Dosage* | *Number of animals* | *Number of rats hypnotized* |
|---|---|---|---|
| Control | 20 ml/kg | 28 | 1 |
| SMS | 11 g/kg | 28 | 6 |
| | 22 g/kg | 28 | 17* |
| Chlorpromazine | 2.5 mg/kg | 28 | 15* |

* $p < 0.01$ when compared with the control

## II. Interactions with inhibitors and stimulants on CNS

### *A. Effect on the hypnotic action of pentobarbital in mice*

One hour after the intragastrical administration of SMS or other drugs, mice were injected intraperitoneally with pentobarbital at a hypnotic dose of 23 mg/kg. The experimentors recorded the number of mice in different drug-pretreated groups that became sleepy within 15 min, as indicated by the disappearance of the righting reflex for more than 1 min. Experimental results from this study indicated that intragastric administration of SMS at a dose of 22 or 33 g/kg or of chlorpromazine at a dose of 2.5 mg/kg could produce a synergistic action with pentobarbital in hypnosis (Table 2.7) (Yuan and Zhang 1990).

### *B. Effect on the stimulant action of chloral hydrate*

One hour after the intragastric administration of SMS or other drugs, animals were given choral hydrate by intraperitoneal injection (190 mg/kg), and the number of mice in different groups becoming sleepy was recorded. The results indicated that when given intragastrically, both SMS (22 or 33 g/kg) and chlorpromazine (4 mg/kg) could produce a synergistic action with chloral hydrate in hypnosis (Table 2.8) (Yuan and Zhang 1990).

### *C. Effect on the stimulant action of deoxy-ephedrine in mice*

Mice given SMS or other drugs intragastrically were immediately injected intraperitoneally with deoxy-ephedrine (2 mg/kg). Forty-five minutes after the injection, activity indices of animals in different groups were detected simultaneously for 10 min by using

*Table 2.8* Synergistic effect of SMS on the action of chloral hydrate

| | *Dosage* | *Number of animals* | *Number of rats hypnotized* |
|---|---|---|---|
| Control | 30 ml/kg | 13 | 1 |
| SMS | 11 g/kg | 13 | 1 |
| | 22 g/kg | 13 | 7* |
| | 33 g/kg | 13 | 8* |
| Chlorpromazine | 4 mg/kg | 13 | 11** |

*, ** $p < 0.05$ and $p < 0.01$ respectively, when compared with the control

*Table 2.9* Antagonistic effect of SMS on the action of deoxy-ephedrine

| | *Number of experimental groups (number of animals)* | *Mobility of mice (mean ± S.D.)* |
|---|---|---|
| Control Solution 25 ml/kg + Saline 10 ml/kg | 8 (16) | 861.6 ± 103.5 |
| Control Solution 25 ml/kg + Metamfetamine 2 ml/kg | 8 (16) | 1512.5 ± 189.5* |
| SMS 16.5 g/kg + Metamfetamine 2 ml/kg | 8 (16) | 1357.0 ± 287.6 |
| SMS 27.5 g/kg + Metamfetamine 2 ml/kg | 8 (16) | 1092.5 ± 114.7† |
| Chlorpromazine 5 mg/kg + Metamfetamine 2 ml/kg | 8 (16) | 787.8 ± 114.7† |

* $p < 0.01$ when compared with the control
† $p < 0.01$ when compared with the analeptic control

the photo-electrical method. Each experiment was consecutively repeated eight times. Results from this study indicated that intragastric administration of SMS at a single dose of 27.5 g/kg could antagonize the deoxy-ephedrine-induced stimulatory action in mice (Table 2.9) (Yuan and Zhang 1990).

### *D. Effect on the stimulant action of amphetamine in mice*

Mice treated intragastrically with SMS or other drugs were immediately injected intraperitoneally with amphetamine (4 mg/kg). One hour after the injection, the activity index of animals was detected for 10 min. Each experiment was repeated consecutively twelve–fourteen times. The results indicated that intragastric administration of SMS at a single dose of 22 g/kg could antagonize the amphetamine-induced stimulatory action in mice (Table 2.10) (Yuan and Zhang 1990).

## III. Interaction with chemical inducers of convulsion in mice

Forty-five minutes after the intragastric treatment with SMS or other drugs, mice were injected intraperitoneally with pentetrazole or strychnine at a dose of 60 or 1.2 mg/kg, respectively. The number of animals showing convulsion was recorded.

*Table 2.10* Antagonistic effect of SMS on the action of amphetamine

| | *Number of experimental groups (number of animals)* | *Mobility of mice (mean ± S.D.)* |
|---|---|---|
| Control Solution 20 ml/kg + Saline 10 ml/kg | 12 (24) | 895.4 ± 135.7 |
| Control Solution 20 ml/kg + Amphetamine 4 mg/kg | 12 (24) | 1258.0 ± 221.3* |
| SMS 11 g/kg + Amphetamine 4 mg/kg | 14 (28) | 1115.1 ± 145.9 |
| SMS 22 g/kg + Amphetamine 4 mg/kg | 14 (28) | 1014.8 ± 115.1† |

* $p < 0.01$ when compared with the control
† $p < 0.05$ when compared with the analeptic control

The results indicated that SMS treatment (33 g/kg) could not prevent the incidence of clonic and tonic convulsion caused by pentetrazole and strychnine, respectively (Yuan and Zhang 1990).

## IV. Body twisting response induced by acetic acid in mice

Forty-five minutes after the intragastric administration of SMS at a single dose of 11 or 22 g/kg, mice were intraperitoneal injected with 0.7 per cent acetic acid (10 ml/kg). The number of body twisting responses of the animals was recorded. The results indicated that SMS could not produce any analgesic action against acetic acid (Yuan and Zhang 1990).

In summary, SMS could significantly reduce the index of voluntary mobility in mice and suppress the fighting and aggressive behaviors of mice living in isolation. SMS could also work synergistically with both barbital (pentobarbital) and non-barbital (chloral hydrate) hypnotic drugs on the CNS. The ensemble of results demonstrated the sedative effect of SMS.

## V. Monoamine neurotransmitters

### *A. Dopaminergic system*

#### *(1) Time dependent changes in the levels of dopamine (DA) and its metabolites (DOPAC and HVA) in rat corpus striatum*

After treating rats orally with SMS at a dose of 11 g/kg, the level of DA in the corpus striatum was elevated, with the maximum level being obtained at 4 hr after the SMS treatment. Then the DA level declined gradually and returned to normal level at 8 hr after the treatment. Meanwhile, DOPAC and HVA (DA metabolites) levels were increased to a maximum in the corpus striatum at 4 hr after the SMS treatment, with DOPAC and HVA levels being increased by about 20 and 40 per cent, respectively. Both DOPAC and HVA levels started to decline thereafter, but they did not return to normal values at 8 hr after the treatment (Table 2.11) (Liu *et al.* 1990).

*Table 2.11* Effects of SMS on the levels of dopamine (DA) and its metabolites in the corpus striatum of rats

| *Time (hr)* | | *Corpus striatum (nμg/g)* | | |
|---|---|---|---|---|
| | | *DA* | *DOPAC* | *HVA* |
| SMS (11 g/kg,ig) | 0 | 7996.0 ± 264.4 | 676.9 ± 34.6 | 642.6 ± 38.5 |
| | 1 | 8130.7 ± 332.1 | 770.4 ± 18.4 | 639.5 ± 52.4 |
| | 2 | 7655.7 ± 210.7 | 898.1 ± 39.6* | 755.4 ± 54.9 |
| | 0 | 8140.6 ± 179.6 | 652.8 ± 39.2 | 623.0 ± 33.4 |
| | 4 | 9011.4 ± 301.7* | 809.2 ± 13.1† | 914.9 ± 62.1** |
| | 8 | 8431.1 ± 335.1 | 738.8 ± 39.4 | 815.4 ± 61.8* |

Each value is mean ± S.E. (n = 6)

* $p < 0.05$ when compared with t = 0
** $p < 0.01$ when compared with t = 0

*Table 2.12* Effects of SMS on the levels of dopamine (DA) and its metabolites in the marginal zone of rats

| *Time (hr)* | | *Peripheral zone (nμg/g)* | | |
|---|---|---|---|---|
| | | *DA* | *DOPAC* | *HVA* |
| SMS (11 g/kg,ig) | 0 | 1231.8 ± 97.1 | 404.7 ± 31.5 | 179.8 ± 23.9 |
| | 1 | 1235.8 ± 88.8 | 402.6 ± 31.4 | 170.1 ± 14.3 |
| | 2 | 951.3 ± 19.2* | 478.5 ± 62.0 | 198.3 ± 23.4 |
| | 0 | 1421.1 ± 72.7 | 420.4 ± 20.5 | 195.2 ± 6.1 |
| | 4 | 1343.6 ± 80.0 | 522.6 ± 33.0* | 229.0 ± 14.4* |
| | 8 | 1352.8 ± 99.4 | 523.4 ± 44.1* | 252.9 ± 10.8** |

Each value is mean ± S.E. (n = 6)

* $p < 0.05$ when compared with t = 0
** $p < 0.01$ when compared with t = 0

### *(2) Time dependent changes in the levels of DA and its metabolites (DOPAC and HVA) in the peripheral zone of rat proencephalon*

In contrast to that in the corpus striatum, the DA level in the peripheral zone of proencephalon was reduced by 20 per cent and then returned to the normal value at 2 and 4 hr respectively, after the oral SMS treatment (11 g/kg). DOPAC and HVA levels were significantly elevated at 4 hr after the treatment, and they showed a continual increase 8 hr thereafter (Table 2.12) (Liu *et al.* 1990).

### *B. 5-Serotoninergic system*

One hour after treating rats orally with SMS at a dose of 11 g/kg, 5-HIAA (a serotonin metabolite) level was elevated gradually in the corpus striatum, with the maximum level being attained at 4 hr after the treatment. The level of 5-HIAA was still significantly higher than the control at 8 hr after the treatment. The rate of elevation of

*Table 2.13* Effects of SMS on the levels of 5-HT metabolites in the corpus striatum and marginal zone of rats

| *Time (hr)* | | *5-HIAA* | |
|---|---|---|---|
| | | *Corpus striatum (nμg/g)* | *Marginal zone (nμg/g)* |
| SMS (11 g/kg,ig) | 0 | 369.0 ± 15.2 | 410.8 ± 25.4 |
| | 1 | 302.0 ± 29.8 | 442.1 ± 33.1 |
| | 2 | 376.2 ± 34.6 | 479.5 ± 29.0 |
| | 0 | 309.8 ± 19.5 | 405.9 ± 26.8 |
| | 4 | 408.7 ± 41.6 | 528.8 ± 29.5* |
| | 8 | 566.8 ± 36.5** | 297.6 ± 35.4* |

Each value is mean ± S.E. (n = 6)
* $p < 0.05$ when compared with t = 0
** $p < 0.01$ when compared with t = 0

*Table 2.14* Effects of SMS on the levels of NA in the peripheral zone and heart of rats

| *Time (hr)* | | *NA* | |
|---|---|---|---|
| | | *Peripheral zone (nμg/g)* | *Heart (nμg/g)* |
| SMS (11 g/kg,ig) | 0 | 271.7 ± 16.5 | 868.8 ± 33.9 |
| | 1 | 265.2 ± 6.6 | 1097.8 ± 77.5** |
| | 2 | 257.7 ± 13.3 | 879.6 ± 56.5 |
| | 0 | 272.4 ± 15.5 | 875.2 ± 106.2 |
| | 4 | 253.7 ± 14.7 | 877.5 ± 70.5 |
| | 8 | 257.0 ± 12.0 | 873.3 ± 58.1 |

Each value is mean ± S.E. (n = 6)
** $p < 0.01$ when compared with t = 0

5-HIAA in the peripheral area was slower than that of the corpus striatum, with a significant elevation of 45 per cent being at 8 hr after the SMS treatment (Table 2.13) (Liu *et al.* 1990).

### C. Noradrenergic system

One hour after the oral treatment with SMS at a dose of 11 g/kg, the noradrenaline (NA) level was significantly elevated in the rat heart, but the effect only lasted for 2 hr, with the NA level being returned to normal. The level of NA in the peripheral zone of proencephalon did not show any significant changes (Table 2.14) (Liu *et al.* 1990).

In summary, oral treatment of rats with SMS (11 g/kg) could significantly decrease the level of DA in the peripheral zone of proencephalon, while the levels of DOPAC and HVA (DA metabolites) as well as 5-HIAA (serotonin metabolite) were significantly elevated. This may be related to the emptying action produced by SMS on monoamine neurotransmitters. Monoamine neurotransmitters were released from the storage site and then metabolized by monoamine oxidase in neurons, both of which were manifested as a decrease in the level of monoamine transmitters and an increase in the levels of their metabolites. The peripheral zone of proencephalon is closely related to human emotional and mental activities. The effect of SMS on this peripheral zone may be related to its sedative effect as well as therapeutic effect on neurasthenia. When compared with the peripheral zone, the effect of SMS on the corpus striatum was different. Oral treatment of SMS could significantly increase the levels of dopamine and its metabolites (DOPAC and HVA) in the corpus striatum. These results indicated that SMS could enhance the dopamine synthesis and metabolism, as well as its exchange rate, resulting in the enhancement of dopaminergic neuron related brain functions. In addition, oral treatment with SMS could significantly increase the level of serotonin metabolite (5-HIAA) in the corpus striatum and the peripheral zone of the rat brain. This may be related to the enhanced exchange rate of serotonin, which is involved in mental activities and endocrine functions. Plausibly, some pharmacological actions produced by SMS may be attributed to its modulatory action in the functioning of the CNS (Liu *et al.* 1990).

## IMMUNE SYSTEM

### I. Phagocytic function of mononuclear phagocyte system (MPS) in mice

According to the carbon particle clearance method of Biozzi (Williams *et al.* 1976), mice were given Indian Ink (tail, iv) at a dose of 0.16 mg/g. At 30 and 300 sec following the administration, a 20 μl blood sample was taken from the orbital vein. The phagocytic index was determined by measuring the absorbance of the blood sample at 675 nm. The results indicated that SMS treatment at a daily intragastric dose of 5.5 or 11.0 g/kg, consecutively for five days, could produce a dose-dependent stimulatory action in mouse MPS, with the effect being similar to that of levamisole when given at daily intraperitoneal dose of 0.01 g/kg (Table 2.15) (Chu *et al.* 1990).

### II. Delayed hypersensitive response in mice

An allergic model was prepared by treating mice with dinitrochlorobenzol (DNCB), and the intensity of delayed hypersensitive response was determined by noting the weight difference between the right ear and the left ear. Both oral treatment (11 g/kg) and intraperitoneal injection (5.5 g/kg) of SMS could produce a very significant stimulatory effect on the delayed hypersensitive response in mice, with the degree of enhancement

*Table 2.15* Effects of SMS on the phagocytic ability of MPS in mice

| | *Dosage (g/kg per day) for 5 days* | *Amount of ink (mg/kg)* | *Number of animals* | *K ± S.D.* |
|---|---|---|---|---|
| Saline | 10.0 | 160 | 20 | 0.0266 ± 0.0012 |
| SMS | 5.5 | 160 | 20 | 0.0492 ± 0.0031* |
| | 11.0 | 160 | 20 | 0.0553 ± 0.0061** |
| Levamisole | 0.01 | 160 | 20 | 0.0642 ± 0.0048*** |

*, **, *** $p < 0.05$, $p < 0.01$ and $p < 0.001$ respectively, when compared with the saline group

*Table 2.16* Effects of SMS on the reaction to DNCB in mouse ear

| | *Dosage (g/kg) and route of administration* | *Number of animals* | *Mean difference in weight between the two ears (mg)* |
|---|---|---|---|
| Saline | 10.0; i.p. | 20 | 32.68 ± 8.22 |
| Levamisole | 0.01; i.p. | 20 | 61.31 ± 15.02* |
| SMS | 5.5; i.p. | 20 | 48.7 ± 10.71*‡ |
| | 11.0; p.o. | 20 | 82.56 ± 19.23*‡ |
| Hydrocortisone-21-sodium succinate | 0.03; i.p. | 20 | 11.91 ± 1.48* |

Each value is mean ± S.D.

* $p < 0.001$ when compared with the saline group
‡ $p < 0.01$ when compared with the levamisole group

*Table 2.17* Effects of SMS on PCA in rats

| | *Dosage (g/kg per day) and route* | *Number of animals* | *Absorbance of Evans Blue exudate from skin plague (mg)* |
|---|---|---|---|
| Saline | 10.0; i.p. | 10 | 0.988 ± 0.09 |
| SMS | 5.5; p.o. | 10 | 0.402 ± 0.05* |
| Disodiumcromoglycate | 0.03; i.p. | 10 | 0.384 ± 0.04* |

Each value is mean ± S.D. (n = 6)

* $p < 0.001$ when compared with the saline group

produced by oral treatment being higher than that produced by intragastric treatment with levamisole (10 mg/kg). However, the extent of stimulatory action produced by the intraperitoneal injection of SMS was significantly lower than that of levamisole (Table 2.16) (Chu *et al.* 1990).

## III. Passive cutaneous anaphylaxis (PCA) in rats

Passive anaphylaxis was induced in rats by administering anti-IgE serum at a titer of 160. The rats were then exposed to pollen (1 per cent in Evans Blue) at a dose of 2.5 mg/ml given intravenously 48 hr later, and then killed 30 min after the injection. The Evans Blue exudates from the skin plaque of different groups of rats were measured for their absorbance. The results indicated that SMS pretreatment (5.5 g/kg/day) could produce a significant inhibitory effect on PCA in rats, with the degree of inhibition being comparable to that produced by the positive control disodiumcromoglycate (0.01 g/kg, i.p.). This suggested that SMS treatment could produce an inhibitory effect on the IgE antibody-mediated immune responses (Table 2.17) (Chu *et al.* 1990).

## ANTIOXIDANT SYSTEM

The ability of SMS to protect against myocardial damage caused by free radicals in various experimental and clinical settings suggests the presence of antioxidant activity

(Lu *et al.* 1994; Rong *et al.* 1989c; Rong *et al.* 1989d). An experimental rat model of carbon tetrachloride-induced hepatic damage was used for detecting the *in vivo* antioxidant activity of SMS, in which oral treatment of SMS at a daily dose of 24 g/kg for 3 days was found to inhibit the carbon tetrachloride-induced damage in rats (Yick *et al.* 1998). The protection was associated with an enhancement on hepatic antioxidant status, particularly the glutathione related antioxidant system (Yick *et al.* 1998). On the other hand, using a rat model of myocardial ischemia reperfusion injury, SMS (24 g/kg/day, 3 days, ig) was shown to be able to protect the heart from free radical-mediated damage, possibly through enhancing the myocardial glutathione antioxidant status (Li *et al.* 1996). In addition, pretreating rats with SMS by direct injection into the duodenum 2 hours prior to cerebral ischemia induced by bilateral carotid artery occlusion suppressed lipid peroxidation and prevented the inhibition of glutathione peroxidase activity associated with ischemia reperfusion (Xuejiang *et al.* 1999). Interestingly, it was found that SMS treatment could produce beneficial effects on the ischemic-reperfused brain tissue even when it was administered 45 minutes after post-ischemic reperfusion. The ability of SMS to non-specifically enhance tissue antioxidant status suggests its preventive effect on aged-related diseases, particularly coronary heart disease and neurological disorders such as Parkinson's disease and Alzheimer's disease, all of which involve free radical-mediated reactions in the pathogenic process.

Using *in vivo* and *in vitro* assay systems, it has been shown that the antioxidant activity of SMS is derived mainly from *Fructus Schisandrae* (FS) (Ko *et al.* 1995c). A novel compound, 5-hydroxymethyl-2-furaldehyde, presumably arising from chemical interaction among chemical constituents derived from FS and *Radix Ophiopogonis* during the decoction process, was also found to possess cardioprotective and antioxidant activity (Yan *et al.* 1998). In an effort to identify the active principle(s) and define the antioxidant mechanism of SMS, the activity-directed fractionation of FS was performed and subsequently obtained a lignan-enriched extract that could enhance hepatic glutathione status in rats (Ko *et al.* 1995b). The beneficial effect of the lignan-enriched FS extract on hepatic glutathione status was evidenced by a generalized protection against hepatotoxicity induced by carbon tetrachloride (Ko *et al.* 1995b), cadmium chloride and aflatoxin (Ip *et al.* 1996). With regard to the molecular mechanism involved in the FS-induced enhancement of hepatic glutathione antioxidant status, experimental results suggested the possible involvement of facilitation of reduced glutathione (GSH) regeneration via the glutathione reductase-catalyzed and NADPH-mediated reactions (Ko *et al.* 1995a). The enhanced GSH regeneration can, in turn, promote the GSH-mediated antioxidant reactions. These findings are consistent with the observation that SMS pretreatment could increase hepatic GSH level as well as glucose-6-phosphate dehydrogenase activity in carbon tetrachloride intoxicated rats (Yick *et al.* 1998). In addition, schisandrin B (Sch B), a dibenzocyclooctadiene derivative isolated from FS, has been found to protect against free radical-mediated hepatic (Ip *et al.* 1995), myocardial (Yim and Ko 1999) as well as cerebral damage (Ko and Lam 2002). The protective effects were associated with the enhancement in tissue glutathione antioxidant status, particularly in the mitochondrion (Ip and Ko 1996). In addition, modulations in tissue level of non-enzymatic antioxidants such as ascorbic acid and $\alpha$-tocopherol, which may be an effect secondary to the enhancement of tissue glutathione status, were also observed (Ip and Ko 1996; Ko and Yiu 2001). Given the tissue non-specific enhancing effect of Sch B on glutathione antioxidant status, a fundamental protective mechanism, such as heat shock protein (Hsp) induction, may be involved in tissue protection. In this regard, a recent study has demonstrated that Sch B pretreatment produced a dose-dependent increase in hepatic Hsp 70 level in mice and protection against tissue necrosis factor-alpha induced hepatic apoptosis in D-galactosamine sensitized mice (Ip *et al.* 2001). In conclusion, Sch B, a major antioxidant constituent present in SMS, can produce tissue non-specific protective effect against oxidative damage by virtue of a yet not clearly defined molecular mechanism. Further works are under way in our laboratory along this line of research.

## OTHERS

### I. Respiration, electrocardiogram and blood pressure

#### *A. Effect on respiration in rats*

When anesthetized rats were given duodenal instillation of SMS at doses of 11 and 44 g/kg, the respiratory frequency was increased by 8 and 12 per cent respectively, indicative of a slight stimulatory action on respiration, 30 min following the treatment. This stimulatory action may be useful for enhancing the respiratory function of the body under shock conditions or in aged people (Zai *et al.* 1990).

#### *B. Effect on electrocardiogram (ECG)*

After the duodenal instillation with SMS at a dose of 11 or 44 g/kg, ECG from the treated animals indicated that there were no detectable changes in the ventricular rate (R-R period) or the time for atrioventricular conduction (P-R period). On the other hand, the T-wave was significantly amplified and the $OT_C$ duration was prolonged. These observations suggested that SMS could produce negative inotropic action with the prolongation of $OT_C$ duration, which is important to the protection against myocardial infarction and arrhythmias (Zai *et al.* 1990).

#### *C. Effect on blood pressure*

Anesthetized rabbits were given SMS by an intravenous injection at doses of 1.1 and 2.2 g/kg. The blood pressure did not show any detectable changes. Rabbits given SMS by duodenal instillation at the same dosage did not show any fluctuations in blood pressure either (Zai *et al.* 1990).

### II. Hemodynamics

An adult dog (8.5–15 kg) was anesthetized by intravenous injection with pentobarbital (30 mg/kg), the chest was then opened to expose the heart. A multi-channel physiological recording and electromagnetic flow meter were used for continual monitoring of the hemodynamic parameters. Intravenous dripping (3.6 ml/min) with SMS at a dose of 1.1 g/kg could produce a negative inotropic effect on the canine heart, which was associated with the decrease in left ventricular internal pressure, myocardial contractile rate and heart rate. The end pressure during left ventricular diastole was found to increase. This may be due to the fact that under negative inotropic status, the

*Table 2.18* Effects of SMS on the enhancement of micronucleus by cyclophosphamide

| | *Number of animals* | *Dosage* | | *Micronucleus frequency (%) (mean ± S.D.)* |
|---|---|---|---|---|
| | | *Cyclophospamide (i.p) (mg/kg per day)* | *SMS (p.o.) (g/kg per day)* | |
| Control | 17 | | | 2.06 ± 1.52 |
| Cyclophosphamide | 17 | 50 × 2 | | 19.35 ± 7.75* |
| Cyclophosphamide + SMS | 18 | 50 × 2 | 14 × 10 | 13.80 ± 7.51† |
| Control | 10 | | | 2.00 ± 1.49 |
| Cyclophosphamide | 8 | 50 × 2 | | 26.20 ± 5.33* |
| Cyclophosphamide + SMS | 7 | 50 × 2 | 28 × 10 | 17.70 ± 10.13 |

* $p < 0.01$ when compared with the control
† $p < 0.05$ when compared with the cyclophosphamide group

cardiac pumping function was impaired, causing the increase in residual blood volume of the left ventricle during diastole, which in turn increased the end pressure during diastole. Results from the experiment also demonstrated that SMS could enhance the cardiac output, coronary flow and vastus arterial flow, but reduce the total peripheral resistance, coronary resistance and vastus arterial resistance. These alterations, which can facilitate the distribution and local supply of blood inside the body, suggested that SMS had the ability to produce a vasodilating effect and reduce blood vessel resistance. On the other hand, lowering of cardiac rate and enhancement of coronary flow, together with the reduced coronary resistance can enable the sufficient perfusion of myocardial tissue, which in turn, further improves the blood circulation (Li *et al.* 1990a).

## III. Tissue protection against toxic agents

### *A. Effect on mutagen-induced chromosomal damage in mice*

Cyclophosphamide given to mice at a daily dose of 50 mg/kg for two days could cause chromosomal damage associated with an enhanced rate of micronucleus formation in bone marrow. Mice treated with SMS orally at a daily dose of 14 g/kg for 10 days could inhibit the cyclophosphamide-induced chromosomal damage, with the rate of micronucleus formation being significantly reduced (Table 2.18).

### *B. Effect on cigarette smoke-induced lung damage in mice*

Mice were smoked by ten or twenty ignited cigarettes for 45 min twice a day (in the morning and afternoon,) for ten consecutive days. Lung tissue of the cigarette smoke-intoxicated mice were taken for tissue sectioning and microscopic examination. Results from microscopic examination indicated that the adhesive membrane in pulmonary bronchioles of control mice were intact, without any upper epithelial proliferation. In addition, no inflammatory cell infiltration was found in the inner and outer wall of bronchioles, and the size of alveolar sac was normal. In contrast, smoke-intoxicated mice had thicker bronchiolar walls, with more lymphocyte infiltration at

*Table 2.19* Elevation of white blood cell count by SMS

| | *Number of animals* | *Dosage* *WBC suppressant (mg/kg per day)* | *SMS (p.o.) (g/kg per day)* | *WBC count (mm$^{-3}$) (mean ± S.D.)* |
|---|---|---|---|---|
| Control | 15 | | | 8480 ± 3195 |
| Cyclophosphamide | 15 | 40 × 5 (i.p.) | | 1990 ± 850** |
| Cyclophosphamide + SMS | 14 | 40 × 5 (i.p.) | 14 × 7 | 3150 ± 1005‡ |
| Hydrocortisone | 14 | 40 × 5 (i.p.) | | 3100 ± 1760* |
| Hydrocortisone + SMS | 14 | 40 × 5 (i.p.) | 14 × 7 | 5896 ± 3287‡ |

* $p < 0.01$ when compared with the control
** $p < 0.001$ when compared with the control
‡ $p < 0.01$ when compared with the cyclophosphamide or hydrocortisone group

the peripheral region, and the upper epithelium of the bronchiolar adhesive membrane had proliferated into a nipple-liked shape. The lung tissue of cigarette smoke intoxicated mice also showed signs of diffuse emphysema. Pulmonary bronchioles isolated from mice treated orally with SMS at a daily dose of 14 g/kg for ten days, was found to be basically normal, without any inflammatory cell infiltration being observed. However, the lung tissue had diffuse emphysema. These observations indicated that SMS could significantly inhibit the inflammation caused by cigarette smoke, but could not improve the condition of pulmonary emphysema caused by cigarette smoke (Hang *et al.* 1990).

### *C. Effect on cyclophosphamide and prednisone-induced leukopenia in mice*

Cyclophosphamide or prednisone given at a daily dose of 40 mg/kg for five days by intraperitoneal or intramuscular injection respectively, could significantly deplete the white blood cell count in mice. Treating mice orally with SMS at a daily dose of 14 g/kg for a week two days prior to the administration of the toxins could significantly prevent the depletion of white blood cell counts caused by cyclophosphamide and prednisone (Table 2.19) (Hang *et al.* 1990).

## IV. Biochemical changes of hepatic tissue in aging rats

Aging rats were treated intragastrically with SMS at a daily dose of 10 g/kg, for three consecutive weeks. Histochemical analysis on hepatic tissue samples indicated that SMS treatment enhanced the response of hepatic sorbitol dehydrogenase, but slightly reduced the responses of both LDH and monoamine oxidase (MAO). The number of glycogen staining responses with intermediate intensity was increased. RNA response was also significantly enhanced. These observations suggested that SMS treatment could increase the hepatic energy metabolism and hence the body vitality, and facilitate the hepatocellular function. All these actions are no doubt beneficial for the improvement of age-related conditions. The relative small effect produced by SMS treatment on LDH and MAO suggested that SMS might not affect the hepatic anaerobic metabolism and the decomposition of monoamines (Zhang *et al.* 1990b).

## CONCLUSIONS

The prescription of multi-component (compound) formulation is a major means for the treatment of diseases in the clinical practice of Chinese Medicine. It is widely accepted that compound formulation, which constitutes one of the crucial components in the wealth of Chinese Medicine, embodies the holistic theory in Chinese Medicine. Clinical practices have long shown that compound formulation is superior to single drug (herb) treatment in regard to its specific therapeutic efficacy. Extensive experimental and clinical investigations on famous traditional Chinese formulations with prudent therapeutic efficacy have important implications in defining the scientific rationale for the theory in Chinese medicine. In addition, these works can also lead to the development of novel Chinese formulation-based pharmaceutical preparations, which maintain the characteristic therapeutic efficacy of compound formulation.

Pharmacological studies of SMS indicate that it can produce effects on the cardiovascular system, central nervous system, immune system, antioxidant system, as well as the ability to resist toxin-induced tissue injury. Given the effect of SMS on the metabolism of monoamine neurotransmitters in the brain and heart of experimental animals, it is likely that the pharmacological effects produced by SMS may, at least in part, be attributed to its integrative action in the CNS and the peripheral effector organs. When taken together, SMS treatment results in the modulation of body functions for the prevention and treatment of diseases.

## REFERENCES


Beijing College of Traditional Chinese Medicine (1973) Effect of SMS on the myocardial ATPase activities in rats and guinea pig, *Xin Yiyaoxue Zazhi*, 10, 387.

Blodel, B., *et al.* (1971) Heart cells in culture: a simple method for increasing the proportion of myoblasts, *Experientia*, 15, 356.

Chu, Y., Cai, S.H., Zhang, L., *et al.* (1990) Shock-arresting and immuno-modulatory effects of SMS. In Y.Q. Yan (ed.), *Integrated Study of Shengmai San Tonic*, CMPSP, China, p.146–60.

Hang, B.Q., Wu, G.Z., Wu, Y., *et al.* (1990) Effect of SMS on lung damage. In Y.Q. Yan (ed.), *Integrated Study of Shengmai San Tonic*, CMPSP, China, p.198–206.

Higgins, T.T.C., *et al.* (1980) The effect of extracellular calcium concentration and Ca antagonist drugs on enzyme release and lactate production by anoxic heart cell cultures, *J. Mol. Cell. Cardiol.*, 12, 909.

Hou, D.H., Li, L., Wang, Q.J., *et al.* (1990) Effects of SMS on experimentally-induced arrhythmia. In Y.Q. Yan (ed.), *Integrated Study of Shengmai San Tonic*, CMPSP, China, p.140–5.

Ip. S.P., Wu, S.S., Poon, M.K.T., *et al.* (1995) Effect of Schisandrin B on hepatic glutathione antioxidant system in mice: Protection against carbon tetrachloride toxicity, *Planta Med.*, 61, 398–401.

Ip, S.P., Ko, K.M. (1996) The crucial antioxidant action of schisandrin B in protecting against carbon tetrachloride hepatotoxicity in mice. A comparative study with butylated hydroxytoluene, *Biochem. Pharmacol.*, 52, 1687–93.

Ip, S.P., Mak, D.H.F., Li, P.C., Poon, M.K.T., *et al.* (1996) Effect of Schisandra Chinensis on aflatoxin B1 and cadmium chloride induced hepatotoxicity in rats, *Pharmacol. Toxicol.*, 78, 413–6.

Ip, S.P., Che, C.T., Kong, Y.C., *et al.* (2001) Effects of schisandrin B pretreatment on tumor necrosis factor-$\alpha$ induced apoptosis and Hsp70 expression in mouse liver, *Cell Stress & Chaperones*, 6, 44–8.

Ko, K.M., Mak, D.H.F., Li, P.C., *et al.* (1995a) Enhancement of hepatic glutathione regeneration capacity by a lignan-enriched extract of Fructus Schisandrae in rats, *Jpn. J. Pharmacol.*, 69, 439–42.

Ko, K.M., Ip., S.P., Poon, M.K.T., *et al.* (1995b) Effect of a lignan-enriched Fructus Schisandrae extract on hepatic glutathione status in rats: Protection against carbon tetrachloride toxicity, *Planta Med.*, 61, 134–7.

Ko, K.M., Yick, P.K., Poon, M.K.T., *et al.* (1995c) Schisandra chinensis-derived antioxidant activities in Sheng Mai San, a compound formulation, in vivo and in vitro, *Phytotherapy Res.*, 9, 203–6.

Ko, K.M., Lam, B.Y.H. (2002) Schisandrin B protects against tert-butylhydroperoxide induced cerebral toxicity by enhancing glutathione antioxidant status in mouse brain Mol. Cell. Biochem. (in press).

Ko, K.M., Yiu, H.Y. (2001) Schisandrin B modulates the ischemia-reperfusion induced changes in non-enzymatic antioxidant levels in isolated-perfused rat hearts. *Mol. Cell. Biochem.*, **220**, 141–7.

Li, H., Hong, S.W., and Peng, Q (1990) Effect of SMS on the hemodynamics of dogs under opened-chest study. In Y.Q. Yan (ed.), *Integrated Study of Shengmai San Tonic*, CMPSP, China, p.166–8.

Li, L.D, Liu, J.X., Wang, R.X., *et al.* (1990a) Effects of SMS on experimentally-induced heart failure in dogs. In Y.Q. Yan (ed.), *Integrated Study of Shengmai San Tonic*, CMPSP, China, p.100–5.

Li, L.D., Sun, H., Gao, F.H., *et al.* (1990b) Effects of SMS on cardiomyocytes under acute heart failure in vitro. In Y.Q. Yan (ed.), *Integrated Study of Shengmai San Tonic*, CMPSP, China, p.128–33.

Li, L.D., Sun, W., Wang, R.X., *et al.* (1990c) Effects of SMS on cardiac ischemia and heart failure in rats. In Y.Q. Yan (ed.), *Integrated Study of Shengmai San Tonic*, CMPSP, China, p.116–27.

Li, P.C., Mak, D.H.F., Poon, M.K.T., *et al.* (1996) Myocardial protective effect of Sheng Mai San (SMS) and lignan-enriched extract of Fructus Schisandrae, in vivo and ex vivo, *Phytomedicine*, **III**, 217–21.

Liu, G.Q., Xie, L., and Liu, X.Q. (1990) Effect of SMS on the monoamine neurotransmitters. In Y.Q. Yan (ed.), *Integrated Study of Shengmai San Tonic*, CMPSP, China, p.235–42.

Liu, Y.G. *et al.* (1978) Effect of Shengmai San (SMS) on the DNA metabolism of cardiac muscle, *Shanxi Xin Yiyao*, 4, 6.

Lu, B.J., Rong, Y.Z., and Zhao, M.H. (1994) Antioxidation effect of SMS on patient with acute myocardial infarction, *Chinese Journal of Integrated Traditional and Western Medicine*, 14, 712–4.

Rong, Y.Z., *et al.* (1983) Ultrastructural changes in adriamycin-induced cardiotoxicosis, *J. Mol. Cell Cardiology*, **15(Supp. 14)**, 30.

Rong, Y.Z., *et al.* (1989) Protection against adriamycin-induced cardiotoxicity by Shengmai San, *Chinese Journal of Cardiology*, 17, 371.

Rong, Y.Z., Wen, W., and Fu, L. (1989) Protective effect of Shengmai San on adriamycin-induced cardiotoxicity – an experimental study, *J. Shanghai Second Medical School*, 3, 39–43.

Rong, Y.Z., Wen, W., and Fu, L. (1989) Protective effect of Shengmai San on adriamycin-induced cardiotoxicity – an experimental study, *J. Shanghai Second Medical School*, 3, 39–43.

Rong, Y.Z., Wen, W., and Yao, M. (1989) Role of oxygen radical in myocardial reoxygenation injury and the protective effect of Shengmai San, *J. Shanghai Second Medical School*, 3, 11–5.

Shi, L., Fan, P.S., Fang, J.X., *et al.* (1990) Effects of SMS on serum lipid and lipoprotein levels in experimentally-induced atherosclerosis. In Y.Q. Yan (ed.), *Integrated Study of Shengmai San Tonic*, CMPSP, China, p.168–79.

Smith, R.L. and Hilker, D.M. (1973) Experimental dietary production of aortic atherosclerosis in Japanese quail, *Atherosclerosis*, 17, 63.

Sun, H., Liu, Z.Y., Gao, F.H., *et al.* (1990) Protective effect of SMS on toxin-induced myocardial injury. In Y.Q. Yan (ed.), *Integrated Study of Shengmai San Tonic*, CMPSP, China, p.180–5.

The Pathophysiology and Anatomy Research Groups of The school of Basic Medical Sciences of Beijing Medical University (1974) A reproducible model of cardiogenic shock in rabbits, *Journal of Beijing Medical University*, 6(4), 241.

The Pathophysiology Research Group of The school of Basic Medical Sciences of Beijing Medical University (1975) Effect of the treatment of cardiogenic shock with SMS in rabbits, *Journal of Beijing Medical University*, 7(2), 118.

Wang, R.X. and Li, L.D. (1990) Scientific bases for the formulation of SMS. In Y.Q. Yan (ed.), *Integrated Study of Shengmai San Tonic*, CMPSP, China, p.12–25.

Williams, C.A., *et al.* (1976) Blood stream clearance of carbon. In *Methods in Immunology and Immunochemistry*, V, Academic Press, New York, p.294.

Xuejiang, W., Magara, T., and Konishi, T. (1999) Prevention and repair of cerebral ischemia-reperfusion injury by Chinese herbal medicine, Shengmai San, in rats, *Free Radical Research*, 31, 449–55.

Yan, Y., Zhu, D., Li, Z., *et al.* (1998) Relationship between component changes and efficacy of Sheng mei san: V. Isolation, identification and content changes of 5-HMF, *Academic Periodic Abstracts of China*, 4, 865–8.

Yick, P.K., Poon, M.K.T., Ip, S.P., *et al.* (1998) In vivo antioxidant mechanism of 'Sheng Mai San', a compound formulation, *Pharmaceutical Biology*, 36, 189–93.

Yim, T.K., Ko, K.M. (1999) Schisandrin B protects against myocardial ischemia-reperfusion injury by enhancing myocardial glutathione antioxidant status. *Mol. Cell. Biochem.*, 196, 151–6.

Yuan, H.N., *et al.* (1985) Introduction to the GJ-series 3-channel mouse activity recorder, *Acta Pharmacologica Sinica*, 1, 49.

Yuan, H.N. and Zhang, Q. (1990) Effect of SMS on the central nervous system. In Y.Q. Yan (ed.), *Integrated Study of Shengmai San Tonic*, CMPSP, China, p.212–8.

Zai, D.Z., Song, P., and Wang, L.Y. (1990) Effects of SMS on the respiratory system, cardiogram and blood pressure. In Y.Q. Yan (ed.), *Integrated Study of Shengmai San Tonic*, CMPSP, China, p.160–5.

Zhang, B.H., Yao, J.A., Zong, Q., *et al.* (1990) Effects of SMS on ischemic heart diseases. In Y.Q. Yan (ed.), *Integrated Study of Shengmai San Tonic*, CMPSP, China, p.133–9.

Zhang, G.X., Liu, Z.Y., Li, C.X., *et al.* (1990) Effect on the chemical composition of the liver in aged rats. In Y.Q. Yan (ed.), *Integrated Study of Shengmai San Tonic*, CMPSP, China, p.206–12.

Zhang, Y.C. (1996), Mechanism of adriamycin-induced cardiomyopathy and its protection, *Ph.D Thesis*, Shanghai Second Medical University.

Zhao, L.G., *et al.* (1974) Prevention of acute heart dysfunction by SMS injection, Tianjin Medical Journal, 2, 449.

Zhuang, S.F., Yan, Y.Q., Zhu, D.N., *et al.* (in press) Effect of SMS and its active ingredients on viral myocaditis.

# 3 Clinical studies on Shengmai San

*Ye-Zhi Rong, Mei-Hua Zhao, Bao-Jing Lu, Xiang-Yang Zhu, Shang-Biao Lu, Ya-Chen Zhang, Jie Chen* [Translated by Kam-Ming Ko]

## CORONARY HEART DISEASE

Shengmai San (SMS), a well-known traditional Chinese medicine (TCM) formula, has been prescribed for replenishing the *Qi* (*vital energy*), recuperating the pulse, rescuing patients from emergency and relieving from collapse. The Ancient Chinese also used SMS for the treatment of '*chest bi-syndrome*' and 'cardialgia'. With the recent advances in medical sciences and technology, particularly in the area of integration of Chinese and modern medicine, the complement of TCM with modern medicine, and vice versa, in terms of theory and practice has allowed the exploration into the wealth of TCM. The clinical applications of SMS have been expanded not only by the introduction of new dosage form, but also by the increased understanding of the mechanisms involved in the prevention and treatment of diseases afforded by SMS treatment (Cong 1980). Nowadays, SMS is widely accepted as one of the effective drugs for the prevention and treatment of coronary heart disease (CHD) in China, particularly for patients with myocardial malfunction (Qiu and Luo 1989). Nevertheless, fundamental differences in the theory between TCM and modern medicine have led to discrepancies in diagnosis and treatment for the same or basically similar diseases. These differences produce inconsistencies between TCM and modern medicine in the definition and nomenclature for certain diseases, like '*chest bi-syndrome*' and 'cardialgia' versus CHD. Hence, the complementary benefits derived from the integration of TCM and modern medicine can definitely facilitate the upgrading of scientific research and therapeutic efficacy of TCM. The investigation of mechanisms involved in the prevention and treatment of CHD by SMS and its clinical applications will be described in the following sections.

### I. Definition and clinical classification of CHD

Coronary heart disease is defined as coronary atherosclerotic heart disease. It is the heart disease induced by myocardial ischemia originating from the blockage of blood vessels due to coronary atherosclerosis. Together with the functional changes of the coronary arteries such as coronary spasm, they are collectively named as CHD.

In general CHD can be diagnosed by electrocardiogram, coronary arteriography, radioactive isotopes and ultrasonic cardiography, laboratory blood tests and clinical manifestations. The World Health Organization has classified CHD into five clinical types: (1) latent; (2) angina pectoris; (3) myocardial infarction; (4) heart failure and arrhythmia and (5) sudden death.

## II. Pathogenesis of CHD

According to modern medicine, the mechanism involved in the pathogenesis of CHD is multifactorial and has not yet been completely defined. Some important pathogenic processes are described as follows.

### *A. Fatty infiltration*

Research studies have demonstrated that the lipid content of atherosclerotic plaque residing in the coronary arterial wall is mainly derived from plasma. Plasma cholesterol, triglycerides and phospholipids are incorporated with apoproteins to form lipoproteins for their dissolution and transportation in blood. The low-density lipoprotein (LDL) mainly consists of cholesterol and cholesterol esters, whereas the very low-density lipoprotein (VLDL) is mainly composed of triglycerides. The increase in plasma lipids arising from the disorder in lipid metabolism can lead to the infiltration of lipids, through the endothelial cells, into the arterial wall. The entry of lipoproteins into the middle layer of the arterial wall will cause the proliferation of smooth muscle cells and their migration towards the arterial intima. The smooth muscle cells and monocytes will then engulf a huge amount of lipids and be converted to foam cells, which, in turn, lead to the formation of a fatty streak, the earliest detectable clinical sign of atherosclerosis. The events, including proliferation of smooth muscle cells, accumulation of connective tissues, disposition of lipids and proliferation of fibrous tissue, finally culminate in the formation of atherosclerotic plaque.

### *B. Aggregation of blood platelets and thrombosis*

Recently, it has been reported that injuries to the arterial intima will cause the adhesion and aggregation of blood platelets at the injured site. The blood platelets then release active agents that can stimulate the proliferation of smooth muscle cells and endothelial cells, as well as the disposition of fibrous proteins. All these eventually lead to the formation of a micro-thrombus at the injured site. The thrombus will then be covered by the newly proliferated endothelial cells and becomes part of the arterial wall. The blood platelets and white blood cells inside the thrombus will be disrupted, with the lipid content being released. These lipids, together with those being deposited from the plasma, eventually lead to the formation of atherosclerotic plaque. Thrombosis without the deposition of plasma lipids will not lead to the formation of atherosclerotic plaque.

### *C. Response to injury*

Hypertension and erratic blood flow created by arterial branching and localized narrowing of blood vessels can produce hemodynamic changes associated with turbulence and shear stress. These hemodynamic disturbances, together with other factors such as the protracted and recurrent noxious effects caused by bacterial and viral infection, toxins, immunogenic factors, vasoactive substances, vasoconstrictive substances and lipid oxidative damage, can produce injuries to the arterial intima as well as functional changes to blood vessels. These, in turn, facilitate the lipid deposition, adhesion and aggregation of blood platelets and finally result in the formation of atherosclerotic plaque.

### *D. Clone theory*

The pathogenesis of atheroclerosis involves the proliferation of smooth muscle cells and their engulfment of lipids. Some growth factors, such as blood platelet-derived growth factor and endothelium derived growth factor, can accelerate both the proliferative process and migration of smooth muscle cells, finally leading to the formation of atherosclerotic plaque.

### *E. Others*

Other mechanisms involved in the pathogenesis of CHD include the changes in neural factors, sex hormones and endothelial function and a decrease in the activity of enzymes located in the arterial wall. Polygenetic inheritance and environmental factors may be also involved.

## III. Historical aspects in the clinical applications of SMS

According to the medical reports from *Ming* and *Ching* Dynasty in China, SMS is capable of benefiting the *Qi* (*vital energy*) and expelling the *heat* in summer. It was therefore prescribed for *heat*-induced damage of the *Qi* and the deficiency of both the *Qi* and the *Yin* caused by the depletion of the *Yin-fluid* (*body fluid*), with symptoms of profuse sweating, thirst, tiredness, shortness of breath, palpitation, weak pulse, and red and dry tongue without saliva. SMS was later used for treating some severe and life-threatening syndromes, like deficiency of *primordial-Qi*, profuse perspiration associated with *Yang* deficiency, coughing and dyspnea due to *lung-asthenia*, and weak and faint pulse conditions. With the rapid advance in the integration of TCM and modern medicine for the prevention and treatment of cardiovascular diseases since the beginning of the 1970s in China, extensive experimental and clinical investigations have been done for SMS. In 1973, Tianjin Nankai Hospital (1973) reported that SMS, when given orally or by intravenous infusion, could significantly elevate the blood pressure and hence protected against the shock state in patients suffering from myocardial infarction. The treatment regimen could also produce positive inotropic action in patients with heart failure. Starting in the late 1970s, some hospitals, particularly those in Beijing and Shanghai, began to use SMS for the treatment of '*chest bi-syndrome*' and 'cardialgia', which is equivalent to angina pectoris and myocardial infarction, respectively. Some hospitals and pharmaceutical manufacturers in Chengdu adopted modern pharmaceutical technologies to prepare dosage form such as intravenous injection liquid or oral liquid for SMS. These preparations were used in animal experiments and clinical studies for investigating the efficacy of SMS. In the mid 1980s, the State Science and Technology Commission (now known as the Ministry of Science and Technology) and the State Drug Administration of China established a nationwide strategic cooperative team for the research and development of SMS. A diverse range of animal studies have been done using SMS oral liquid, and the clinical efficacy on angina pectoris and cardiomyopathy as well as on retarding the aging process have also been examined. In 1992, the State Chinese Medicine Authority regarded SMS injection liquid as the first batch of essential medication in the emergency room of TCM hospitals. In 1994, the State Administration of Traditional Chinese Medicine, China, established a cooperative team for emergent '*chest bi-syndrome*,' which coordinated all clinical studies from six hospitals in Beijing, Shanghai and Chengdu on evaluation of clinical efficacy of SMS on angina pectoris as well as the related animal studies. In 1995, a comprehensive report was made for 219 case studies on angina pectoris and other related studies on SMS, and the overall result was encouraging (SATCM, China, 1995).

Nowadays, the scope of indications for SMS has been widened. SMS can be used not only for the treatment of cardiovascular diseases, such as CHD (angina pectoris and myocardial infarction), heart malfunction, cardiogenic shock, arrhythmia, sick sinus syndrome, myocarditis and cardiomyopathy, but also for improving conditions of the *Qi* and *Yin* deficiency caused by heart surgery, cardiopulmonary diseases, septic shock, epidemic hemorrhagic fever and cancer chemotherapy. In addition, SMS can also produce anti-aging effects.

## IV. Experimental studies and clinical applications of SMS

In the past three decades, a huge amount of experimental evidence concerning SMS has been accumulated. Currently, SMS is clinically used for the treatment of both *Qi* and *Yin* deficiency associated with various cardiovascular diseases. The better understanding of the mechanisms involved in the pharmacological actions of SMS has enabled a more favorable clinical outcome of SMS treatment in CHD to be achieved. Some of the important findings are described as follows.

### *A. Free radical scavenging and anti-lipid peroxidation effects: prevention and treatment of myocardial ischemia-reperfusion injury*

Recently, free radical-mediated reactions and lipid peroxidative damage have been implicated in the pathogenesis and development of CHD (Yao *et al.* 1988; Huang *et al.* 1993). Free radicals are involved in the formation of atherosclerotic plaque through oxidatively modifying the LDL. The abrupt increase in lipid peroxidative products during the atherosclerotic process can damage the arterial endothelial cells, which will, in turn, accelerate the progress of atherosclerosis. Meanwhile, the lipid peroxidative reactions can alter the lipid components of membrane bound enzymes, receptors and ion channels, leading to functional impairment of these proteins. Transmembrane lipid peroxidation can produce new ion channels, which are highly permeable to calcium ions, resulting in intracellular calcium overload. Superoxide radical ($O_2^{\bullet-}$) and hydroxyl radical ($^{\bullet}OH$) are reactive oxidants that can indirectly or directly cause oxidative chain reactions in membrane lipids, resulting in the formation of malondialdehyde (MDA), one of the end-products of lipid peroxidation. Under normal conditions, the cellular antioxidant defense system can protect against free radical-mediated oxidative damage. Superoxide dismutase (SOD) and glutathione peroxidase (GSH-Px), the important enzymatic components of the antioxidant defense system, can prevent the formation of $^{\bullet}OH$ by removing $O_2^{\bullet-}$ and $H_2O_2$ as well as terminate the propagation of lipid peroxidation reactions by converting the oxidized lipid molecules into non-oxidative hydroxyl compounds.

A comparative study between 98 cases of CHD and 110 healthy individuals, as described by Lu *et al.* (1994), has demonstrated that SOD and GSH-Px activities were significantly lower and MDA level was significantly higher in the blood of CHD patients. The decrease in GSH-Px activity and the increase in MDA level were more apparent in patients with myocardial infarction (68 cases), when compared with those with angina pectoris (30 cases). These observations suggested that under conditions of myocardial infarction, the increased extent of myocardial ischemia could enhance the free radical production and hence the lipid peroxidation reactions. The free radical-mediated lipid peroxidation reaction is an important factor involved in the pathogenesis and development of CHD (Lu *et al.* 1994). In the same study, patients with myocardial infarction were randomly divided into groups: (1) standard treatment and (2) standard treatment plus SMS treatment. SMS was given orally in the form of granules, with 3 packs per day (one pack contains *Radix Ginseng* (1 g), *Radix Ophiopogonis* (3 g) and *Fructus Schisandrae* (1.5 g)), consecutively for 28 days. The results indicated that after 2 weeks of standard treatment plus SMS, the patients were found to have higher blood SOD and GSH-Px activities than those of patients receiving only the standard treatment *per se*, with significant increase in enzyme activities being observed four weeks after the SMS treatment (Lu *et al.* 1994). These findings indicated that SMS treatment could enhance antioxidant defense and inhibit lipid peroxidation in patients suffering from myocardial infarction. However, the mechanism involved in the beneficial effect produced by SMS treatment remains unclear. Research studies on individual herbs indicated that saponin isolated from *Ginseng* could suppress the free radical production from activated polymorphonuclear leukocytes and neutrophils, while lignans isolated from *Schisandrae* could reduce the extent of lipid peroxidation.

During the procedure of open-heart surgery, a myocardial tissue sample was taken from the post-ischemic heart after 30-min of reperfusion for microscopic examination. It was found that myocardial tissue obtained from patients treated with SMS prior to surgery were found to have a smaller degree of ultrastructural damage when compared with the control. Plasma creatine phosphokinase isozyme (CPK-MB) activity in SMS-treated patients was also lower than that of the control 12 and 24 hours after the surgery (Li *et al.* 1994).

### B. Effects on myocardial energy metabolism, intolerance to ischemia and infarct size

A huge volume of clinical data indicated that the decrease in exercise-tolerance is not only directly correlated with the occurrence of cardiogenic shock in patients suffering from CHD, but also predictive to an impaired quality of life during the course of the disease (Deng and Luo 1992; Pathophysiological Research Group, Beijing Medical University 1977). Exercise-tolerance or exercise capacity is a factor independent of changes in the ST segment of the electrocardiogram or the extent of pathological changes in the coronary artery that affects the survival of a patient with CHD. It is usually expressed as maximal oxygen consumption ($VO_{2max}$). $VO_{2max}$ indicates the largest capacity of the body in the delivery and hence utilization of oxygen under exercise conditions. A study conducted in Xin-Hua Hospital has examined the short-term effect of SMS injection liquid on the exercise-tolerance of patients suffering from angina pectoris. Twenty-seven patients with angina pectoris, capable of doing flat-board exercise, were randomly divided into the SMS pretreated group and Danseng (*Radix Salviae Miltiorrhizae*) pretreated group. SMS or Danseng pretreatment was done by intravenous infusion at a daily dose of 80 and 16 ml, respectively, consecutively for fourteen days. After the drug pretreatment, both groups were assessed for the ability to perform flat-board exercise and the maximal oxygen consumption was computed as metabolic capacity. The results indicated that patients given SMS injection liquid were found to have significantly increased $VO_{2max}$ when compared with the Danseng group. The electrocardiogram of patients given with SMS injection liquid was also significantly improved, as indicated by the prolongation in the consecutive exercise period and the time delay for the suppression of the ST segment by 1 mm. The recovery time for the ST segment and the absolute value at maximum suppression of the ST segment were also decreased. These observations indicated that SMS injection liquid was superior to that of Danseng, as assessed by exercise-tolerance and the extent of myocardial ischemia in patients with angina pectoris. As described in Li *et al.* (1999), patients suffering from angina pectoris, as diagnosed by coronary arteriography, were subjected to a flat-board exercise test, before and after the treatment with SMS injection liquid. The results indicated that pretreatment with SMS was able to improve heart function, enhance exercise-tolerance, reduce the total sum of ST segment suppression measured at various leads, and ameliorate the extent of myocardial ischemia in patients suffering from angina pectoris under exercise conditions (Li *et al.* 1999).

Numerous clinical studies on the therapeutic effect of SMS preparations on acute myocardial infarction have been reported. Regardless of the presence or absence of shock symptom, the therapeutic application of SMS could significantly reduce mortality caused by the acute myocardial infarction (Zhang *et al.* 1984). A study from the No.1 Affiliated Hospital of Zhongshan Medical University showed that when SMS was used for treating 35 cases of myocardial infarction, only one patient died and two patients had recurrent infarction in a three-year post-treatment surveillance. It has been reported that when treating elder patients suffering from acute myocardial infarction with SMS in concomitant with a western drug, the effective rate was 93.8 per cent, which was significantly higher than that afforded by the western drug alone *per se* (82.0 per cent) (Hua and Qi 1996). While the mortality rate of SMS-treated patients was 6.2 per cent, it was 18.0 per cent in patients given the western drug *per se* (Hua and Qi 1996). By using the Swan-Ganz catheter, the hemodynamic effect produced by SMS injection liquid on patients with acute myocardial infarction was examined in Xiyuan Hospital (Dong *et al.* 1984). The result indicated that the infusion of 15–30 ml of SMS (10 ml/min) into the right atrium could increase the stroke volume 5 min after infusion, with the peak value being attained after 10–15 min and maintained for another 35 min. On average the stroke volume was increased by 19.5 per cent. Patients given SMS showed a flushed complexion, reduced sweating and warmness of limbs (Dong *et al.* 1984). A study from Tianjin Nankai Hospital (1973) has shown that when patients suffering from acute myocardial infarction associated with shock were given a western drug treatment, 7 out of 13 patients died. When the western medication was supplemented with SMS injection liquid or *Si Ni Tang* (a decoction for treating *Yang* exhaustion), none of the 10 patients died. As reported by Mo *et al.* (1997), SMS, *Astragali*, and urokinase were administered for thrombolytic treatment of acute myocardial infarction. While there was no observable difference in the incidence of coronary reopening, when compared with the urokinase treatment group, SMS, when administered together with urokinase, could reduce the incidence of post-thrombolysis chest pain, complications of myocardial infarction, and the level of plasma angiotensin II, with the state of hypoxemia being improved (Mo *et al.* 1997). It has also been reported that SMS injection liquid produced significant therapeutic efficacy on the atrioventricular block associated with acute myocardial infarction (Zhu and Qiao 1998).

### *C. Effects on heart contractility, cardiac output, coronary flow and heart function*

Clinical manifestations of CHD like shortness of breath, lassitude and palpitation are directly related with the heart function. The treatment with SMS injection liquid could significantly ameliorate these symptoms (Fang *et al.* 1987). Experimental results indicated that SMS could enhance heart contractility and left ventricular function. In this regard, the component herbs in SMS, namely *Ginseng* and *Schisandrae*, were found to produce positive inotropic action (Shi *et al.* 1981). Various *Ginseng*-related preparations could increase the contractility of isolated frog hearts and intact hearts in rabbits, cats, and dogs, with the effect being independent of atropine and the influence of vagus nerves (Shi *et al.* 1981).

It is now believed that the mechanism involved in the positive inotropic actions of SMS is related to its ability to suppress the $Na^+/K^+$ ATPase activity of cardiomyocytes, with the mode of action being similar to that of inotropic saponins. As observed by Qin *et al.* (1983), SMS injection liquid inhibited the $Na^+/K^+$ ATPase activity of rat cardiomyocytes by 25.0 per cent, whereas an ionic fluid, containing equivalent amounts of $Na^+$, $K^+$, $Ca^{2+}$, and $Mg^{2+}$ present in SMS, inhibited $Na^+/K^+$ ATPase activity by only 6.5 per cent. While the enhancement in cardiac output afforded by positive inotropic drugs is usually associated with an increase in peripheral resistance, the positive inotropic effect produced by SMS is accompanied by the reduction in peripheral vascular resistance to a varying degree. Therefore, SMS can increase the flow-output, but not the pressure-output, with a reduction in oxygen consumption. An electromagnetic flow-meter was catheterized to the ascending branch of the left circumflex coronary artery in anesthetized dogs, the coronary blood flow was found to increase following the intravenous infusion of SMS injection liquid at a dose of 1 ml/kg (Zhang and Yang 1986). The increased coronary flow was also associated with a concomitant enhancement in the peripheral arterial flow (Zhang and Yang 1986). Zhou *et al.* (1997) used a doppler ultrasonic detection technique and reported that in the absence of β-receptor blocker and calcium antagonist, SMS injection liquid was able to improve the left ventricular diastolic function. The result suggested that SMS injection fluid could be used for the treatment of uncomplicated left ventricular diastolic dysfunction or both diastolic and systolic dysfunction. The therapeutic action of SMS is therefore different from that of conventional positive inotropic drugs (Zhou *et al.* 1997).

Clinical studies have demonstrated that SMS treatment could reduce the incidence of angina pectoris and the extent of myocardial damage caused by acute myocardial infarction, and facilitate the repair of damaged myocardial tissue (Wang *et al.* 1997). SMS treatment could also reduce the incidence of complications and mortality caused by acute myocardial infarction. Because of the faster onset of therapeutic action with an injection of SMS when compared to an oral treatment, patients under critical conditions with life-threatening symptoms are usually administered with SMS injection liquid (every 10 ml of SMS injection liquid is comprised of *Radix Ginseng* (1 g), *Radix Ophiopogonis* (3 g) and *Fructus Schisandrae* (1.5 g)). The Shanghai City Clinical Cooperative team for SMS Injection Liquid, as led by the Department of Cardiology in Xinhua Hospital (1999), has reported the clinical effect of SMS injection liquid on 248 patients suffering angina pectoris. All patients selected for the study fulfilled the diagnostic criteria of the World Health Organization (WHO) for coronary heart disease-derived angina pectoris published in 1979. SMS injection liquid was given by intravenous infusion at a daily dose of 80 ml, consecutively for fourteen days, while the control group was given Danseng (*Radix Salviae Miltiorrhizae*) injection liquid at a daily dosage of 16 ml. With reference to the evaluation criteria established by the National Conference on the Integration of Chinese and Western Medicine for the Prevention and Treatment of Coronary Heart Disease, Angina Pectoris and Arrhythmia in 1979, the respective treatment would be regarded as effective if the symptom of angina pectoris was reduced by more than one grade and the use of nitroglycerin was not required or reduced by more than 50 per cent in dosage (Department of Cardiology, Xinhua Hospital 1999). In addition, if the ST segment was increased by more than 0.05mV after the drug treatment under resting conditions, the electrocardiogram would be regarded as being improved. The treatment would be regarded as ineffective if the symptomatic improvement was not obvious or stable. Results indicated that SMS injection liquid was superior to Danseng injection liquid in therapeutic efficacy on relieving the shortness of breath, myocardial ischemia-induced chest pain (chest distress) and palpitation (cardiac terror), with the improvement in symptoms being observed on the third to fifth day after the drug treatment. As assessed by doppler ultrasonic measurement, the cardiac output was found to be increased, and the peripheral vascular resistance decreased, both were indicative of improved heart function. The effective rate, as assessed by electrocardiogram, was 71.2 per cent for the SMS group, which was better than that of the Danseng group (60.0 per cent). Meanwhile, SMS treatment could produce a similar action to Danseng in decreasing the blood viscosity. The Cooperative Team for Emergent Treatment of *Chest bi-syndrome* Under the State Administration of Traditional Chinese Medicine, China (SATCM 1995) conducted a clinical investigation in which 219 patients suffering from angina pectoris associated with deficiency of both *Qi* and *Yin* were treated with SMS injection liquid by intravenous infusion at a daily dose of 40 ml. Results indicated that the effective rate in relieving pain was 95.0 per cent, and the effective rate in ameliorating other symptoms was 93.6 per cent. The effective rate in stopping and reducing the use of nitroglyceride drugs was 61.5 per cent, and that in improving the electrocardiogram under ischemia was 68.5 per cent. The myocardial oxygen consumption was also significantly reduced. SMS injection liquid tended to produce a better therapeutic efficacy on unstable angina pectoris (Sun and Wang 1998). By *differentiation of signs and symptoms*, the therapeutic efficacy afforded by SMS injection liquid on angina pectoris of deficiency of both *Qi* and *Yin* type was superior to that of the *heart-blood stasis* type. It has been reported that the total effective rate of SMS injection liquid was 92.5 and 78 per cent, respectively, for unstable angina pectoris (Sun and Wang 1998) and angina pectoris (Wang *et al.* 1997) unresponsive to diltiazem and nitroglycerin treatment. The effective rate in improving the electrocardiogram for asymptomatic myocardial ischemia was 93.8 per cent (Su 1998).

### D. Effect on thrombosis and blood clotting function

CHD is usually associated with an enhancement in a series of blood adhesive factors. On the incidence of angina pectoris or at the early stage of myocardial infarction, a variety of cells like vascular endothelial cells, blood platelets, and leukocytes will release a large amount of vasoactive substances including angiotensin II, thromboxane, and platelet activating factors (PAF). Up until now, PAF is the strongest known stimulant for the platelet activation and aggregation, which can facilitate thrombosis by increasing the blood viscosity and reducing the blood flow.

Experimental studies from the Shanxi Provincial Academy of Traditional Chinese Medicine demonstrated that SMS injection liquid could inhibit the process of thrombosis in rabbits, with the time of thrombus formation being prolonged and the mass of the thrombus being reduced (Xu *et al.* 1986). SMS injection liquid also significantly prolonged the prothrombin time and time of plasma prothrombin consumption, as well as reducing the fibrinogen content of the thrombus. The results suggest that SMS injection liquid could ameliorate the high tendency for blood coagulation. Lu *et al.* (1998) have examined the changes in blood viscosity in 53 cases of CHD before and after the SMS injection liquid treatment (80 ml/day). The results indicated that SMS treatment significantly reduced the blood viscosity (including the upper and lower shear limit of the whole blood, plasma viscosity and red blood cell sedimentation), which was associated with a reduction in the adhesion, as well as the 1-min, 5-min, and maximal aggregation of blood platelets. The action mechanism may be related to the ability of Shengmai injections to enhance the synthesis of prostacyclin, inhibit the production of thromboxane, and selectively antagonize the PAF.

### *E. Modulatory effect on blood pressure and improvement in microcirculation*

SMS can produce a biphasic modulatory action on blood pressure, with the systolic blood pressure being slowly increased under the condition of hypotension and being slightly decreased under the condition of hypertension, while the normal blood pressure is not affected (Pathophysiology Research Group, Beijing Medical University 1978). It has been reported by Li (1978) that SMS injection liquid was used for treating six patients of acute myocardial infarction accompanied by cardiogenic shock. Four of the patients, who failed in the treatment with vasopressive amine, were switched to the treatment with SMS injection liquid, while the other 2 patients were only given the SMS treatment throughout the study. SMS treatment was found to be efficacious, with blood pressure of all treated patients being increased by 20 mmHg on average. The onset of pressive effect being observed 15–20 min after intravenous infusion, which was different from instantaneous action produced by vasopressive amine. In addition, the elevated blood pressure in SMS-treated patients could be sustained for at least 7 hours, indicating that the SMS has a modulatory effect produced by SMS treatment on vascular tone. However, the mechanism involved remains to be elucidated (Li 1978).

### *F. Others*

It has been reported that SMS could increase the cholesterol level of high-density lipoprotein, but did not affect the level of total cholesterol and triglyceride in patients suffering from CHD. In addition, SMS could modulate the level of endogenous glucocorticoid hormone. It has also been reported that SMS was capable of significantly reducing the level of ß-receptor-induced cAMP level in the patients with CHD. Plasma levels of anti-coagulase and $PGF_{1\alpha}$ were increased, while the concentration of thromboxane $B_2$ was decreased in CHD patients treated with SMS (Wang *et al.* 1999). SMS treatment could also increase the rate of myocardial DNA synthesis (Liu *et al.* 1978).

## V. Conclusions

From experimental studies to clinical applications for the prevention and therapy of CHD, the results of SMS treatment have been encouraging. Given the absence of or little side effect produced by naturally-occurring herbal formula, the clinical application of SMS for the treatment of cardiovascular diseases, particularly the prevention and treatment of CHD, is very promising. As TCM is gaining attention worldwide, experimental and clinical investigations on SMS and its related preparations will proceed further in the future. The future trend in research and development of SMS should be focused on the following aspects: 1) the further elucidation of dose-dependent relationship; 2) the extraction and the subsequent recombination of major active components from the herbs and 3) in-depth investigation on the action mechanism of SMS in cardiovascular system from cellular and biochemical level to molecular and gene level.

# CARDIAC ARRHYTHMIA

Cardiac arrhythmia (also called arrhythmia), an abnormal rhythm of heart activities, can occur in patients suffering from various kinds of organic heart diseases and other diseases. The clinical significance of arrhythmia is determined by the etiology, duration and hemodynamic changes associated with the arrhythmia. Severe arrhythmia can cause the decrease in cardiac output, insufficient organ perfusion and even death. Clinical manifestations include palpitation, chest-pain, chest distress (the feeling of oppression over the chest), shortness of breath, syncope and sudden death.

Arrhythmia is classified into two main types, namely tachycardia and bradycardia. The more commonly occurring tachycardia is associated with incidences of atrial fibrillation, atrial flutter, supraventricular paroxysmal tachycardia, and ventricular tachycardia, while sick sinus syndrome and interruption could be found in the cases of bradycardia.

The earliest clinical record of arrhythmia in traditional Chinese medicine can be dated back to the period of the *Han* Dynasty. Dr. Zhang Zhong-jing had described the symptoms of and therapy for the conditions of irregular pulse in '*Treatise on Exogenous Febrile Diseases*'. He also reported that the slow and irregular pulse and palpitation could be treated with *Glycyrrhizae* decoction. After the *Tsui* and *Tang* Dynasty, reports on pulse studies gradually emerged. The '*Pinhu's Sphygmology*' by Li Shi-zhen had described the pulse conditions of arrhythmia in further detail, like slow pulse, irregular pulse, and abrupt pulse. Studies from ancient physicians in TCM have accumulated a huge volume of information for the understanding and clinical treatment of arrhythmia.

SMS is a TCM formula prescribed for benefiting the *vital energy* and recuperating the pulse. Nearly one thousand years ago, SMS had already been used for the treatment of arrhythmia. Not until the recent decades, the therapeutic action of SMS has further been confirmed by clinical investigations. The therapeutic effect of SMS on tachycardia and bradycardia will be described in the following sections.

## I. Treatment of tachycardia with SMS

### *A. Atrial fibrillation*

Atrial fibrillation is a manifestation of irregular contraction of atrial cardiac muscles, with the frequency being 400–600 beats/min. Atrial fibrillation is usually associated with common organic heart diseases like coronary heart disease and rheumatic heart disease. In addition, it can also be found in patients suffering from hyperthyroidism. Idiopathic atrial fibrillation can be found in 5–6 per cent of patients who have no history of any heart diseases.

Symptoms of atrial fibrillation, including palpitation, shortness of breath and contractile failure, are related to the change in heart function and ventricular rate. The detachment of a thrombus from the wall of the vessel lumen can lead to myocardial infarction.

### *B. Treatment of atrial fibrillation with SMS*

According to the theory of TCM, the pathogenesis of atrial fibrillation caused by rheumatic heart diseases usually involves the deficiency of both *heart-Qi (vital energy)* and *heart-Yin*, and the inability to disperse of *water* and *dampness*. The respective therapy should therefore be aimed at replenishing the *vital energy* and nourishing the *Yin* for rectifying the abnormal body function (i.e. cause of the disease) as well as inducing diuresis for symptomatic relief. As such, SMS injection liquid (80 ml, supplemented with 500 ml of physiological liquid adjuvant) was given by an intravenous drip once a day for 2 weeks (Zhu *et al.* 1997). Meanwhile, digitalis and diuretics were prescribed for reducing cardiac workload and improving heart contractile function. The pathogenesis of atrial fibrillation caused by coronary heart disease usually involves the deficiency of *heart-Qi* and the blockage of *vessels* and *channels* (*meridian*). The corresponding treatment should also be aimed at replenishing the *Qi* (for removing the cause of the disease) and opening the *vessels* and *channels* (for relieving the symptoms). SMS could be used for benefiting the *Qi* and at the same time, vasodilators and anticoagulants were prescribed.

The therapeutic strategy for atrial fibrillation should integrate the *Differentiation of Signs and Symptoms* (TCM) and clinical diagnosis (modern medicine). The kind of treatment would primarily be determined by the course of the disease and the cardiac rate. Therapeutic interventions from modern medicine should be adopted for the rapid control of over-driven atrial fibrillation associated with serious hemodynamic disorders. When treating atrial fibrillation without obvious hemodynamic disorders, the adoption of digitalis based treatment together with SMS usually produced good therapeutic outcome.

### *C. Premature beat*

Premature beat is advanced heart beating caused by the earlier impulse generated from ectopic focus. This focus can be of atrial, junctional, and ventricular types according to its location. Atrial ectopic focus is second to the more commonly occurring ventricular focus. Premature beat, which can be found in normal individuals, is associated with *psychasthenia*, tiredness, cigarette smoking, and alcoholic intake. It can also be caused by pathological changes in the mitral valve, myocarditis, cardiomyopathy, cardiopulmonary diseases and congenital heart diseases. Ventricular premature beat can also be found in normal individuals; however, it is more likely associated with organic heart diseases like coronary heart disease, myocarditis, cardiomyopathy, and mitral valve prolapse.

TCM regards premature beat as a result of emotional upset, stagnation of *liver-Qi*, deficiency of *heart-Qi* and exhaustion of *heart-Yin*. The insufficient *heart-Yang* results in water retention and the *heart-blood stasis* associated with severe palpitation.

While single or occasional incidences of premature beat would not produce any symptoms, some of the affected individuals may have discontinuous pacing or intermittent feeling of strong and forceful beating. Frequent or persistent premature beat can reduce cardiac output and the blood perfusion of major organs, causing palpitation, lassitude, angina pectoris or breathlessness.

### *D. Treatment of premature beat with SMS*

Given that the integration of TCM and modern medical treatment can produce a more effective preventive and therapeutic outcome, the therapeutic strategy for premature beat should be based on both the *Differentiation of Signs and Symptoms* (TCM) and clinical diagnosis (modern medicine). As described by Wang *et al.* (1997), SMS injection liquid was used for treating 46 cases of coronary heart disease and hypertension associated with positive ventricular late potential (VLP) by intravenous infusion at a single daily dose of 40 ml, for 15 consecutive days. The results indicated that the rate of negative VLP was 82.6 per cent, with the ventricular and atrial premature beat disappearing in all VLP negative individuals. As reported by Chen and Wu (1996), the integration of TCM and modern medicine was adopted in treating twenty-two cases of refractory ventricular arrhythmia. It was found that 82 per cent of the treated cases were effective, particularly in arrhythmia caused by early coronary heart disease, viral myocarditis, and other unknown causes. The blood pressure and cardiac rate did not show any significant change before and after the drug treatment. The effect of a supplemented SMS preparation on patients suffering from coronary heart disease and showing positive VLP has been reported by Zeng (1996). Thirty-two cases of positive VLP were randomly divided into two groups, which were given supplemented SMS preparation and 'heart painkiller', respectively. The patients of the two groups did not have any significant differences in ages, sex, and the course of sickness. Results from the study indicated that the rate of negative VLP afforded by the supplemented SMS preparation treatment was 75.0 per cent, while that afforded by the 'heart painkiller' was 9.38 per cent, with the group difference being statistically significant ($p < 0.01$). The author believed that the supplemented SMS preparation could significantly protect against myocardial ischemia/reperfusion-induced injury and improve the metabolism of cardiomyocytes. It can also nullify the unsynchronized depolarization and delay the conduction of bioelectricity in cardiomyocytes. All of these actions produced by the treatment of supplemented SMS preparation could eventually lead to the conversion of coronary heart disease-associated VLP from positive to negative sign. The use of a supplemented SMS preparation is also effective for the prevention of lethal arrhythmia. By the *Differentiation of Signs and Symptoms* in TCM, arrhythmia can be classified into three types, namely deficiency of *Qi* and *Yin*, stagnation of *Qi* and *blood stasis*, and stagnation of *phlegm-dampness*. Patients of these three types of arrhythmia were treated with corresponding herbal decoction including SMS and performed *Qigong*-respiratory exercise. Significant therapeutic efficacy of these treatment regimens was found, with total effective rate being 86 per cent. The result indicated that the present treatment regimen comprising herbal decoction and Qigong exercise was an effective method for treating arrhythmia. As reported by Zhou (1996), twelve cases of frequent premature beat, with the course of the disease being longer than three months and failed in anti-arrhythmic drug treatments, were treated with amiodarone together with a high dose of oryzanol and *Astragali* supplemented SMS oral liquid. Results indicated that the effective rate was 83.3 per cent, with ten and two cases being cured and failed, respectively. The author suggested that these drugs may exert their therapeutic effect indirectly through metabolic regulation or protection against ischemia. Through the intermediacy of the central nervous system or peripheral nervous system, the favorable changes in hemodynamics and/or metabolism can produce the therapeutic effect. In this case, the three drugs, namely amiodarone, *Astragali* and SMS, may act synergistically in producing the therapeutic effect, which cannot be afforded by the individual use of the drugs. Du *et al.* (1998) have investigated the effect of SMS injection liquid on the arrhythmia caused by the complication of hyperthyroidism. Forty-three cases of hyperthyroidism with frequent atrial premature beat and ventricular premature beat were divided into treatment group and control group. Treatment group (22 cases) was given 10 mg of methimazole orally, three times per day, together with SMS injection liquid (20 ml) (in 500 ml physiological liquid adjuvant containing 5 per cent of glucose) via intravenous infusion, once per day. The control group (21 cases) was only given methimazole at the same dosage regimen used in the treatment group. Results indicated that after two weeks of drug treatment, the effective rate in protecting against arrhythmia was 72.7 and 47.8 per cent in treatment group and control group, respectively, with the difference between the two groups being statistically significant ($p < 0.05$). The results suggested that SMS injection liquid could be used for treating arrhythmia in patients suffering from hyperthyroidism.

## II. Treatment of bradycardia with SMS

Bradycardia, a heart rate slower than 60 beats per minute, is usually associated with sinus bradycardia, sick sinus syndrome, sinus arrest, sinoatrial block, and atrioventricular block. Its incidence is likely related to pathological changes in the myocardium, increased vagal influence on a normal pacemaker, hyperkalemia, and side effects of certain drugs.

The descriptions of bradycardia in TCM include palpitation, severe palpitation, *chest bi-syndrome* (pain), dizziness, *Jue*-syndrome and the slow and irregular pulse conditions. The disorder was believed to be caused by the deficiency of *Yang-Qi* and the *stasis* of *heart-blood.*

### *A. Clinical manifestations*

Mild cases of bradycardia produce no symptoms, or only the feeling of oppression over the chest and palpitation. But severe cases are associated with symptoms such as dizziness, shortness of breath, lassitude, syncope, and Adams-Stokes syndrome.

### *B. SMS treatment*

When the installation of an artificial pacemaker in patients suffering from sick sinus syndrome is not feasible, the treatment with SMS plays a significant role in the clinical management of bradycardia. The administration of a combination of SMS (for replenishing the *Qi* and nourishing the *Yin*) and a herbal formula (*Bao Yuan Tang*, for preserving the *primordial-Qi*) was found to be more effective than SMS alone, in enhancing cardiac rate and improving heart function.

As described by Zhu *et al.* (1995), the clinical efficacy of SMS (when administered together with nicotinamide) for treating senile bradycardia was examined. One hundred and sixty eight cases were given SMS orally and nicotinamide by intravenous infusion. Results indicated that the total effective rate was 99.0, 93.8 and 87.5 per cent, in treating sinus bradycardia, sick sinus syndrome and heart block respectively, indicating that the combined use of SMS with nicotinamide could produce synergistic action in shortening the course of disease and enhancing the therapeutic efficacy.

## PRIMARY CARDIOMYOPATHY

Primary cardiomyopathy, a disease state originated from the myocardium, is clinically classified into idiopathic cardiomyopathy and specific cardiomyopathy. Idiopathic cardiomyopathy is the cardiomyopathy associated with impaired heart functions, and can be further classified into dilated cardiomyopathy, hypertrophic cardiomyopathy, restrictive cardiomyopathy, and arrhythmogenic right ventricular cardiomyopathy. Specific cardiomyopathy is associated with specific heart diseases or other systemic diseases, which include ischemic cardiomyopathy, valvular cardiomyopathy and hypertensive cardiomyopathy, etc.

### I. Treatment of dilated cardiomyopathy-derived heart failure by Shengmai San

Since there is no curative treatment for dilated cardiomyopathy, symptomatic treatment is mainly used at the present time. While the majority of cases of dilated cardiomyopathy were not diagnosed until the initial strike of heart failure, the treatment regimen is generally similar to that of treating heart failure, with the aims being to 1) relieve symptoms and thereby improve the quality of living; and 2) to protect the myocardium in order to prolong the survival time. For mild cases of heart failure, the treatment is mainly focused on protecting the myocardium, while the amelioration of symptoms and protection of the myocardium are more critically needed in severe cases of heart failure. If all these treatments remain ineffective, cardiac transplantation would be required.

#### *A. Improvement in left ventricular functions and exercise-endurance*

Dilated cardiomyopathy is the chronic deterioration of pump functions caused by diffused myocardial damage associated with systolic or diastolic disorder of the left ventricle or bilateral ventricles. Its causes could be idiopathic, familial or hereditary, viral and/or immunological, due to alcoholic or toxic factors, as well as secondary to known cardiovascular conditions. In clinical situations, there has not been a single ideal drug for the disease.

Patients with myocardiopathy given SMS treatment showed significant improvement in left ventricular systolic functions, as assessed by M-type echocardiogram and systolic time intervals. Jiang *et al.* (1988) have reported that oral SMS treatment could reduce the internal circumference of the left ventricle and significantly increase the ejection volume indices (EF and CO), suggesting that SMS could improve the deteriorated left ventricular systolic functions. Since the improvement in ventricular functions was not associated with changes in cardiac rate and blood pressure, it was postulated that the effect was mediated by an enhancement in myocardial contractility. In this connection, SMS treatment was shown to improve the circumferential fiber-shortening rate, which could more directly reflect the myocardial contractile function. This observation further indicated that SMS could enhance the myocardial contractility under conditions of cardiomyopathy.

### *B. Effect of SMS on exercise-eudurance and its implications*

It is well accepted that the efficacy of drugs used for treating chronic heart malfunction can be most sensitively assessed by an endurance study using physical exercise. While one of the targets in the management of cardiomyopathy is to enhance the 'quality of living' of the patients, an endurance study provides a convenient and reproducible means of assessing the effectiveness of treatments. Studies by Jiang *et al.* (1988) demonstrated that cardiomyopathic patients receiving SMS treatment showed enhanced endurance to physical exercise, without an accompanying increase in oxygen consumption (indicated indirectly by multiplying the heart rate with the systolic pressure). These observations suggested that SMS treatment could improve cardiac mobility without increasing the myocardial oxygen consumption in the patients of cardiomyopathy. A huge amount of reports have indicated that SMS and its herbal component, *Ginseng*, could enhance the tolerance of the myocardium towards anoxic conditions and protect the myocardial ultrastructure from damage, thereby enhancing the endurance to physical exercise. All in all, SMS may produce multiple effects for the improvement of endurance under conditions of cardiomyopathy.

### *C. Improvement in the hemodynamic markers by SMS injection liquid under conditions of heart failure induced by dilated cardiomyopathy*

Dilated cardiomyopathy is a common form of cardiomyopathy manifested by bilateral ventricular expansion, reduction in ventricular contractility and cardiac output, and increased ventricular end-diastolic pressure. Patients usually die of heart failure, arrhythmia or sudden death. No effective drug is available for the treatment of dilated cardiomyopathy. Even with the use of inotropic saponin (cardiac glycosides), diuretics, and vasodilator, their prognosis is still unfavorable. Recent studies have indicated that SMS, which can improve the left ventricular function, has been prescribed for coronary heart disease, acute myocardial infarction and toxic shock.

Zhang *et al.* (1990) have studied the hemodynamic effect of Shengmai injection liquid on the patients of dilated cardiomyopathy with heart failure (22 cases) by recording their bifunctional echocardiograms. Results showed that after the intravenous injection of SMS, the stroke volume (SV), cardiac output (CO), cardiac index (CI), ejection fraction (EF), endocardial fractional shortening ($\Delta$D per cent), and thickening rate of the ventricular wall ($\Delta$T per cent) were significantly increased, while the systemic vascular resistance (SVR) was significantly decreased. The heart rate, average arterial pressure, E-point septal separation (EPSS), and the end-systolic stress (ESS) did not show any significant changes. The effects of SMS on the pathophysiology of the dilated cardiomyopathy were manifested by the expansion of bilateral ventricle, attenuation of ventricular contractility, reduction in cardiac output, and the enhancement in ventricular end-diastolic pressure (Zhang *et al.* 1990).

Patients exhibited a declined pump function even in the asymptomatic compensation period, but the deterioration of pump functional markers becomes significant as symptoms appear. Clinical observation indicated that the stroke volume, cardiac output, cardiac index, ejection fraction, endocardial fractional shortening, and thickening rate of ventricular wall were significantly lowered in patients of dilated cardiomyopathy. After intravenous infusion of SMS preparation, various parameters that reflect ventricular contractility were significantly elevated, suggesting the improvement in cardiac pump function. Among these, elevations in the shortening rate of the left ventricular axis ($\Delta$D per cent) and thickening rate of the ventricular wall ($\Delta$T per cent) were particularly significant. The myocardial fiber during contraction of left ventricle was shortened by 60 per cent, the shortening percentage change was closely correlated with ejection fraction, and is usually greater than 30 per cent. Ventricular wall thickening during the diastolic period is an important manifestation of myocardial contractile functions. Echocardiography examinations demonstrated that the ventricular wall mobility of the patients was generally lowered, with the thickening rate of the ventricular wall ($\Delta$T per cent) being significantly decreased. After the drug treatment, contractility of the left posterior ventricular wall and the intraventricular septum was enhanced, as indicated by the significant increases in shortening rate of the right ventricular axis ($\Delta$D per cent) and thickening rate of the ventricular wall ($\Delta$T per cent) (Zhang *et al.* 1990). These observations have suggested that SMS injection liquid may enhance myocardial contractility and cardiac output, which are consistent with its known pharmacological actions.

E-point septal separation (EPSS) is a sensitive marker for assessing right ventricular function. When heart function is impaired, early diastolic bicuspid blood flow and the extent of opening of bicuspid anterior lobe would be reduced. On the other hand, the EPSS value would increase, usually not more than 5–8 mm, and show a negative and positive correlation with the ejection fraction and end-diastolic pressure, respectively. Patients of dilated cardiomyopathy were found to have significantly elevated EPSS values. After SMS injection, the cardiac output was significantly enhanced while the EPSS did not show any significant changes. The failure of SMS to change EPSS may be due to the relatively weak heart function in the patients, leading to the small expansion in left ventricular internal circumference.

### D. *Effect of SMS injection on the after-load of dilated cardiomyopathy*

During the course of heart failure in dilated cardiomyopathy, the left ventricular after-load would increase as a result of decreased cardiac output and autonomic vasoconstriction of the systemic circulation, resulting in a further decrease in cardiac output. The increase in after-load would usually cause myocardial hypertrophy. However, under the condition of dilated cardiomyopathy, the increase in myocardial weight was not accompanied with thickening of the ventricular wall, despite an overloading of the ventricle. The ratio of ventricular wall thickness and internal circumference of the heart chambers tends to be normal or decreased, without causing compensatory cardiac hypertrophy.

Studies indicated that during the compensatory stage of heart function, patients with dilated cardiomyopathy were found to have reduced myocardial contractility, lowered ejection fractions, but a normal peripheral resistance in the systemic circulation. When the heart functions further deteriorated with incidence of apparent heart failure, EF would reduce to less than 40 per cent, with a concomitant increase in peripheral vascular resistance. This abnormal increase in the after-load, termed 'after-load dismatch', can significantly impair the contractile function of the heart, and it is one of the important factors for the induction of heart failure, in which the use of vasodilators can improve the heart functions.

Studies with dogs indicated that injection of *Ginseng* saponins could reduce the vascular resistance in vertebral and vastus arteries. In patients, treatment with Shengmai injection liquid could lower the PVR and the left ventricular after-load, thereby elevating the CO, CI, and EF. These findings point to a direct vasodilating action of SMS on resistant vessels. Given its ability in enhancing myocardial contractility and dilating blood vessels, SMS injection liquid may be a candidate for the effective treatment of heart failure in dilated cardiomyopathy.

## II. Combination of Chinese medicine and western medicine treatment for the intractable heart failure of the dilated cardiomyopathy

### *A. Combined use of SMS and diuretics*

Dilated cardiomyopathy is usually associated with intractable heart failure, particularly an uncontrollable heart rate. While digitalis treatment can easily produce poisoning effects, treatment with high dosages of β-receptor blockers may have the side effects of inducing or exaggerating heart failure, so these drugs cannot effectively improve the heart functions. The use of SMS in adjunct with western drug treatments can avoid these shortcomings of using a western drug alone, and result in good therapeutic efficacy (Yuan and Zheng 1997).

By the *Differentiation of Signs and Symptoms* in traditional Chinese medicine, dilated cardiomyopathy belongs to mild or severe 'palpitation', manifested by deficiencies of the *Qi* (*vital energy*) and *Yin*, and retention of *dampness* and anxiety (Zhang H. 1997). These symptoms should therefore be treated by invigorating the *Qi* and nourishing the *Yin*, eliminating *dampness* and diuresis, as well as tranquilizing the mind. Modern pharmacological studies have shown that SMS could improve heart function and enhance the myocardial contractility, with its positive ionotropic action being similar to that of lanatoside. SMS can also enhance the oxygen inhaling and carrying capacity of the body, thereby invigorating the *Qi*.

The combined use of SMS and diuretics can reduce cardiac workload, enhance the myocardial contractility, relieve heart failure, and improve heart functions. Yuan and Zheng (1997) reported the study in 100 cases of intractable heart failure of dilated cardiomyopathy with a co-treatment of SMS and diuretics. Apparent therapeutic efficacy was found, as indicated by the significant improvement in various heart functional markers, including cardiac output (CO), stroke volume (SV), cardiac index (CI), ejection fraction (EF), and total peripheral resistance (TPR). These observations indicated that the combined use of SMS and diuretics was superior to the conventional single drug treatment for treating the intractable heart failure of dilated cardiomyopathy.

### *B. Combined use of SMS and dopamine*

Lu *et al.* (1997) described the use of SMS injection liquid in combination with dopamine in 34 cases of heart failure associated with dilated cardiomyopathy. SMS treatment had a significant effect in 15 cases, and was effective in another 16 cases, with an overall effective rate of 91.2 per cent.

In addition to its anti-arrhythmic effects, SMS injection could improve the heart functions by fundamental means, such as enhancing the myocardial tissue metabolism and facilitating the restoration of deteriorated myocardial functions, indicating that SMS is an effective drug for heart failure.

Dopamine is a non-glycoside inotropic drug, and is the precursor for the biosynthesis of noradrenaline and adrenaline. It can activate the $\alpha$- and $\beta$-adrenergic receptors, as well as the dopamine receptors. It has been reported that the use of an intravenous infusion of dopamine in treating level IV heart dysfunction of different pathogenic origins, except in the case of *wind* epigastralgia, produced a satisfactory therapeutic effect. At a normal dosage, dopamine could enhance myocardial contractility and cardiac output, induce arterial/venous vasodilation, and improve microcirculation. It could also enhance renal blood flow and glomerular filtration rate as well as facilitate urination and sodium excretion, thereby reducing the workload of the heart. Dopamine could also reduce the vascular resistance of pulmonary circulation, and increase the blood supply to the myocardium by stimulating coronary vasorelaxation, thereby producing beneficial hemodynamic changes. At higher dosages, it could significantly increase both systolic pressure and arterial pressure. While the short-term use of dopamine is highly effective, drug tolerance develops as the body becomes desensitized after prolonged use, and as a result the efficacy of the drug declines. On the other hand, SMS injection liquid produces the actions of invigorating the *Qi*, nourishing the *Yin*, strengthening the pulse, replenishing exhaustion, inducing vasodilation, and enhancing coronary flow. The co-administration of SMS and dopamine may suggest an effective treatment for heart failure in dilated cardiomyopathy by virtue of their synergistic action that can shorten the course of treatment and minimize the incidence of drug tolerance.

### *C. Combined use of SMS and dobutamine*

Tan (1997) reported the use of SMS injection in combination with dobutamine for treating heart failure in dilated cardiomyopathy (35 cases) and observed good therapeutic efficacy. The significantly effective rate for clinical symptoms/signs and the total effective rate were found to be 60 and 90 per cent, respectively. The heart-chest ratio, the left ventricular end-diastolic internal circumference and the incidence of arrhythmia were significantly reduced.

The theory of traditional Chinese medicine holds that somatic *asthenia* is critical in the pathogenesis of dilated cardiomyopathy, particularly under the condition of *Qi*-deficiency. The *Qi* is the *vital energy* for the body. Deficiency in *primordial-Qi* could impair visceral functions, causing metabolic inadequacy associated with *Qi*- and *Yin*-deficiencies, and malnourishment of the *heart*. Deficiency of the *Qi* could impair blood circulation, manifested as a weak pulse, *blood stasis* and palpitations. Cardiomyopathy is usually asymptomatic, incurable and recursive in nature, causing damage to the *spleen* and *stomach* eventually. While the deficiency of the *spleen* would hinder the transformation of *body fluid*, deficiency of the *kidney* would lead to the accumulation of *dampness*, causing edema and retention of fluid in the *lung*, shortness of breath with coughing, and disturbance in renal reabsorption and dyspnea when prostrate.

*Ginseng* is the '*Monarch*' herb (the principal herb) in the formulation of SMS. It produces actions of invigorating the *lung*, benefiting the *Qi* through the *channels* to the *lung*, *spleen* and *heart*, invigorating the *primordial-Qi*, invigorating the *spleen* and nourishing the *lung*, promoting *body fluid* production and tranquilizing the mind. *Ophiopogoniae*, being the '*Minister*' herb in SMS, exerts its actions through the *channels* to the *lung*, *stomach* and *heart*, nourishes the *Yin*, benefits the *stomach*, moistens the *lung*, and dissipates *heat* in the *heart*. *Schisandrae* serves as the '*Assistant*' herb that acts through the *channels* to the *lung*, *heart*, and *kidney* in benefiting the *Qi* and promoting *body fluid* production, invigorating the *kidney* and nourishing the *heart*, restoring the astringent functions, retaining the *lung-Qi*, and preventing excessive perspiration. The invigorating, astringent and dissipating actions of *Ginseng*, *Schisandrae* and *Ophiopogoniae*, respectively, act synergistically in SMS to supplement the *Qi* and nourish the *Yin*, to promote *body fluid* production and prevent thirst, to retain the *Yin* and restore the astringent functions, to prevent collapse by enriching the *Qi*, and recuperating normal pulse and *body fluid* content. For the treatment of dilated cardiomyopathy, SMS can restore organ functions and strengthen the healthy *Qi*. Pharmacological studies indicate that SMS is capable of 1) producing positive inotropic actions; 2) enhancing the coronary flow; 3) improving myocardial metabolism; 4) modulating the blood pressure; 5) improving the microcirculation; 6) inhibiting the process of thrombosis; 7) reducing oxygen consumption of myocardium; and 8) modulating the level of endogenous glucocorticoid hormone. Therefore, SMS treatment could enhance the myocardial contractility and improve the heart functions of patients with cardiomyopathy.

Dobutamine can activate the ß1-, ß2- and α-adrenergic receptors. Since its stimulatory action on the ß1-receptors is much stronger than that on the ß2-receptors, it can increase the myocardial contractility and cardiac output. Meanwhile, its stimulation on the ß2-receptors results in the dilation of the peripheral vessels, thereby lowering the peripheral vascular resistance and the pulmonary wedge pressure. Therefore, dobutamine can produce definite therapeutic effects on intractable heart failure, provided that it is not administered at high dosages. Worsening of heart functions, such as elevated heart rate and increased peripheral vascular resistance, has been reported when dobutamine was given at doses exceeding 10 μg/kg/min. For the treatment of dilated cardiomyopathy, the dosage of dobutamine should be carefully controlled. A low dose of dobutamine (5 μg/kg/min) co-administered with SMS injection liquid has been reported to produce apparent efficacy.

On the other hand, dobutamine has been reported to lower the levels of potassium and magnesium in blood, thereby disturbing the electrolyte balance and exaggerating ventricular arrhythmia. In contrast, when dobutamine was co-administered with SMS injection liquid, the occurrence of these side effects has been significantly reduced. It is obvious that anti-arrhythmic action of SMS is favorable to the treatment of dilated cardiomyopathy, which is usually associated with arrhythmia. In fact, the co-administration of SMS and dobutamine could significantly reduce the occurrence of ventricular arrhythmia and reverse cardiac enlargement, as assessed by the heart-chest ratio and the end-diastolic internal circumference of the left ventricle. The restoration of heart functions by co-administering SMS and dobutamine may implicate their synergistic action in reducing heart workload, improving myocardial metabolism, and reducing left ventricular end-diastolic pressure.

## III. Treatment of arrhythmia of dilated cardiomyopathy by SMS injection liquid

The incidence of frequent ventricular premature beat and discontinuous ventricular tachycardia associated with dilated cardiomyopathy are 70–95 and 40–80 per cent, respectively. The relationship of tachycardia with sudden death has yet to be determined. Although asymptomatic discontinuous ventricular tachycardia is usually observed in the ambulatory electrocardiogram, the use of anti-arrhythmic drugs for the treatment of dilated cadiomyopathy remains questionable.

Dilated cardiomyopathy is usually associated with various types of arrhythmia, and its treatment regimen is usually subjected to different limitations. Zhang *et al.* (1997) studied the effect of SMS injection liquid in 40 cases of dilated cardiomyopathy complicated with arrhythmia. After one week of SMS treatment, gastrointestinal symptoms like nausea and vomiting disappeared in 38 cases, while symptoms like palpitation and shortness of breath disappeared or were significantly relieved after one course of treatment. Among 23 cases with ventricular premature beat, 16 of these had symptoms that disappeared after one course of treatment without using any anti-arrhythmic drugs; 4 cases with atrial fibrillation did not recover, but in these the heart rate was reduced; and 24 cases of ventricular blockade did not show any significant change. Among 36 cases that had lowered ST segment, inverted or flattened T-wave, 18 cases had their ST-T restored to normal after one course of treatment.

## IV. Treatment of anemic cardiomyopathy

Severe anemia, with a hemoglobin level lower than 50 g/L, can induce myocardial anoxia and myocardial fatty degeneration. The resultant myocardial hypertrophy and cardiac dilatation in association with an ultimate decrease in myocardial reserve function, as manifested by a high cardiac output cardiomyopathy or even congestive heart failure, is refractory for treatment. Congestive heart failure is frequently present in clinical situations and accounts for the major cause of death in macro-erythroblastic anemia. Chen and Chen (1996) reported the use of SMS and dopamine in resuscitating an anemic cardiomyopathic patient from acute heart failure induced by ventricular overload during blood transfusion.

Manifestation of cardiogenic shock induced by congestive heart failure resembles that of the *depletion* syndrome in traditional Chinese medicine. *Depletion* can be divided into *Yin*- and *Yang-depletion*. *Yang depletion* is manifested as coma, unconsciousness, cold sweat, pale complexion, cyanotic lips, clammy limbs, incontinence, pale tongue with thin fur, and weak or irregular pulse. The *Ginseng-Aconiti* Decoction is commonly prescribed for the treatment of *Yang depletion* that should be tackled by restoring the *Yang*, tonifying the *Qi* and preventing collapse. On the other hand, *Yin depletion* is manifested as reddish and dry skin with slight sweating, general fever and dryness of the mouth, weak and rapid pulse, reddish tongue with yellow fur. Treatment should focus on benefiting the *Qi*, nourishing the *Yin*, and strengthening the pulse by astringent, in which SMS is commonly used.

The manifestation of heart failure induced by anemia generally resembles that of *Yin depletion*, with signs of *Yang*-deficiency of the *heart*. With regard to its treatment, SMS can be used as the '*Monarch*', and *Radix Aconiti* Preparata as the '*Minister*', to facilitate resuscitation and prevention of collapse. According to the theory of traditional Chinese medicine, the herbal components of SMS, namely *Ginseng, Ophiopogoniae* and *Schisandrae* have *Qi*-invigorating, *Yin*-nourishing and tranquilizing effect, respectively, while *Aconiti Preparata* has resuscitating effect. Results from recent studies indicated that SMS was capable of 1) improving left ventricular function, enhancing myocardial contractility and endurance in physical exercise; 2) increasing the ejection fraction of the left ventricle and blood pressure, reversing mild and moderate shock, reducing the dosage requirement of hypertensive drugs for treating severe shock, and reducing the dependence on dopaminergic drugs; 3) decreasing the plasma content of β-globulin, platelet Factor IV and $TXB_2$, inhibiting the response of platelet activation and release; 4) reducing the permeability of blood capillaries; 5) enhancing systemic tolerance to anoxia; and 6) reducing the area of myocardial infarction, prolonging the beating time of ischemic heart and improving the microcirculation under conditions of cardiogenic shock. After all, effective management of refractory heart failure induced by anemic cardiomyopathy requires early diagnosis and prompt treatment.

## V. Treatment of apical hypertrophic cardiomyopathy by SMS

Apical hypertrophic cardiomyopathy, a subtype of hypertrophic cardiomyopathy, is a genetic disease usually found in males, with clinical manifestations being similar to that of other types of hypertrophic cardiomyopathy. Two main diagnostic properties are described: 1) amplified inverted T-wave (>10 mm) and QRS-complex with high voltage (Rv5 > 26 mm or Rv5 + Sv1 > 35 mm) on the electrocardiogram; and 2) concentric hypertrophy of cardiac apex observable in the left ventricular cardiograph and 2-dimensional echocardiogram, with the apical thickness ($24.8 \pm 6.6$ mm) being far larger than normal ($9.4 \pm 3.1$ mm). Ratio of apical thickness and the thickness in mid-region of left ventricular anterior wall is significantly enhanced ($1.86 \pm 0.53$) when compared with normal ($1.05 \pm 0.24$). There are only a few studies on the treatment of apical hypertrophic cardiomyopathy. Some reported that nifedipine and propranolol treatment could produce a certain degree of efficacy, with the relief of chest pain and enhancement of endurance in physical exercise. While some studies reported that verapamil treatment could reduce the occurrence of amplified T-wave, others reported that such treatment could not produce significant improvement in the electrocardiogram.

Qiu and Luo (1989) described a case of apical hypertrophic cardiomyopathy given nifedipine and diuretics treatment. Although the symptoms of chest distress and palpitation were relieved, this treatment regimen could not prevent the occurrence of arrhythmia, leaving the patient at risk of sudden death. Since the case was complicated with the presence of sinus bradycardia, the use of anti-arrhythmic drugs was limited. In addition, the patient showed intensive response to the side effects of many drugs, thereby rendering pharmaceutical intervention ineffective. In contrast, satisfactory therapeutic effects have been observed with SMS treatment. With slightly more than a year of management after discharge from the hospital, the health status of the patient has returned to normal: heart functions have been improved, incidence of arrhythmia basically disappeared, and general mobility was enhanced. All these findings suggested the therapeutic effect of SMS towards hypertrophic cardiomyopathy.

## VIRAL MYOCARDITIS

Myocarditis, inflammatory pathological changes of the myocardium, is mainly caused by an infection of bacteria, viruses, fungi, and protozoa. A small number of cases are related to toxic response or intoxification, radiation exposure, and systemic diseases like dermatomyositis and sarcoidosis. In recent years, the incidence of rheumatic fever or bacteria (e.g. diphtheria)-induced myocarditis has progressively decreased, whereas viral myocarditis has become more prevalent. The latter is mostly found in young patients, but the occurrence in children, middle-aged and elder people is not rare. Viral myocarditis has posed a big threat to community health that can hamper the productivity of the society. While there is still no specific treatment available for viral myocarditis, the treatment based on the integration of Chinese and modern medicine has obtained some successes. The following describes the use of SMS in the prevention and treatment of viral myocarditis.

### I. Pathogenesis

According to modern medicine, the pathogenesis of viral myocarditis involves (1) the direct effect of viral invasion on the myocardium, causing damage on the microvasculature and (2) indirect damage to the myocardium caused by the immune responses associated with the disease. According to the theory of TCM, viral myocarditis belongs to the clinical aspects of terrified palpitation, severe palpitation, cardiac obstruction, and consumptive diseases. The pathogenesis of the disease involves the deficiency of the *primordial-Qi* and pathogen invasion, while emotional status, tiredness, and exogenous infection are factors leading to the induction of the disease. The disease can damage the *Qi*, *blood*, *Yang* and *Yin* not only to the *heart*, but also to the whole body (Sun 1995).

### II. Clinical diagnosis

There is still some controversy about the diagnostic criteria for viral myocarditis. Under clinical conditions, there are no rapid and highly specific diagnostic means available for viral myocarditis. According to the National Forum on Myocarditis and Myocardial Diseases held in 1987 (Editorial 1987), the diagnostic criteria for viral myocarditis include: (1) clinical symptoms and signs (upper respiratory tract or digestive tract) of viral infection; (2) heart performance at the acute stage or within 1–3 weeks after viral infection, persistent and frequent pulses unrelated to feverish condition, edema, coughing, non-productive asthma, chest distress, heart malfunctions with unknown causes, and other signs such as cardiomegaly, cardiac murmur, the third and fourth heart sound and arrhythmia; and (3) various arrhythmias found in the electrocardiogram: (a) atrioventricular block or sinoatrial block, bundle branch block, (b) more than two channels with its ST segment being lowered or elevated, or abnormal Q concentration, (c) frequent, multiform, polyfocal or paired/parallel premature beat, burst or paroxysmal supraventricular tachycardia and ventricular tachycardia, beating and fibrillation, (d) frequent atrial or ventricular premature beat, (e) inverted T-wave (either bipolar or flattened). In addition, other indications of viral myocarditis include positive response of PCR EVS-RNA test and a four-fold increase in antibodies against Coxsackie B virus (Zhao *et al.* 1996), positive response of serum creatine phosphokinase (CK) and its isozyme (CK-MB), and positive response of serum myocardial troponin T (cTnT) and troponin I (cTnI). With regard to sensitivity and specificity, serum cTnI is superior to cTnT as a diagnostic for acute viral myocarditis. In addition, cTnI also provides a wider time window for diagnosis. Hence, serum cTnI is a highly recommended diagnostic marker for viral myocarditis in clinical situations. Finally, the diagnosis of viral myocarditis should be carefully distinguished from other pathological conditions, such as hyperthyroidism, β-adrenoreceptors hyperfunction syndrome, coronary heart disease, rheumatic myocarditis, toxic myocarditis, and primary/secondary cardiomyopathy, that can also affect the myocardium.

## III. Clinical treatment: effect of SMS

There is no specific drug treatment available for acute viral myocarditis. The patients of viral myocarditis should have plenty of bed rest and a sufficient nutrient intake. With the adoption of the therapeutic method integrating modern medicine and TCM, the recovery time for the patients is about three months (Editorial 1987). Patients with unstable pathogenic conditions should have a longer resting time. In general, the treatment regimen for viral myocarditis consists of the following.

### *A. Anti-infection treatment*

For western drug treatment, ribavirin is usually used. Other anti-bacterial drugs are given simultaneously if necessary. Regarding TCM, Double Coptidic decoction, Jade Screen powder, *Radix Sophorae Flavescentis*, and *Lonicerae-Forsythiae* powder are the drugs of choice (Chen *et al.* 1990; Sun 1998; Wang 1998). Experimental studies have demonstrated that *Radix Sophorae Flavescentis* was effective against Coxsackie virus, while *Radix Astragali*, SMS, and *Calcui Bovis* acid can enhance the immune functions in patients suffering from viral myocarditis (Jia 1998; Xiong *et al.* 1997). SMS can benefit the *vital energy* (*Qi*) and nourish the *Yin*, and also produces a modulatory effect on the cell-mediated immune function. SMS treatment can enhance the immune function of cardiomyocytes and the peripheral vascular sub-populations of T-lymphocytes, thus improving the immune status of cardiomyocytes under the condition of viral myocarditis.

### *B. Myocardial nutrients*

Co-enzyme A (Co-A), co-enzyme $Q_{10}$ (Co-$Q_{10}$), adenosine-triphosphate (ATP), fructose-1,6-diphosphate (FDP), and vitamins can be used. In acute treatment, ATP (40 mg) and Co-A (200 U) supplemented to glucose (5 per cent) solutions (250 ml) with or without FDP (10 g) are given by intravenous infusion once a day, with the course of treatment lasting for 10–14 days, while Co-$Q_{10}$ and vitamins are given orally.

### *C. Symptoms-oriented treatment*

Patients with heart failure or arrhythmia are given standard drug treatment. If necessary, the restoration of cardiac rhythm by direct current is given in case of severe intractable tachycardia (supraventricular tachycardia and ventricular tachycardia). Specific treatment is not required for patients with I°/II° atrioventricular block or if the ventricular

*Table 3.1* Comparison between the SMS treatment group and the placebo control group in heart function

| | | *Stroke volume (ml/beat)* | *Cardiac output (ml/min)* | *Cardiac index* |
|---|---|---|---|---|
| SMS | Before (n = 20) | 60.6 ± 14.9 | 4.4 ± 1.0 | 2.5 ± 0.5 |
| | After (n = 20) | 72.6 ± 16.2* | 5.3 ± 0.9 | 3.1 ± 0.6* |
| Control | Before (n = 20) | 66.1 ± 8.9 | 4.7 ± 0.6 | 2.9 ± 3.8 |
| | After (n = 20) | 62.3 ± 10.5 | 4.0 ± 1.2 | 2.8 ± 0.5 |

Values given are mean ± SD, with the number of cases indicated in parentheses.

* significantly different from that prior to the SMS treatment ($p < 0.001$)

rate is more than 50 per minute. If the symptoms are obvious and the cardiac rate is less than 50/min, the patients are given either atropine or isoproterenol orally, at a dosage of 0.3 and 10 mg respectively, 3 times a day. And if necessary, isoproterenol (0.5–1.0 mg) supplemented with glucose (5 per cent) solution can be given by intravenous infusion. Under conditions of II° and III° of Mobitz type II atrioventricular block, a broad QRS complex being observed on the electrocardiogram or the recurrent incidence of Adams-Stokes syndrome, the patients should be temporarily installed with an artificial pacemaker.

### *D. Corticosteroid treatment*

Whether corticosteroid hormones can be used to treat viral myocarditis is still a controversial area. They are not used as a routine or standard treatment regimen, but in severe cases it will be used as early as possible, for protecting the cardiomyocytes and reducing the edema. In contrast, it should not be used at the early stage (within 10 days) of the course of general viral myocarditis.

Based on the findings from our previous studies on the effect of SMS on cardiomyopathy and the aforementioned treatment approaches, clinical investigation of the effect of SMS treatment on viral myocarditis has been done in the early 1990s. In this clinical study, 35 patients with viral myocarditis were given SMS preparation (1 package is comprised of *Radix Ginseng Destillata* (1 g), *Radix Ophiopogonis* (3 g) and *Fructus Schisandrae* (1.5 g); provided by The First Chinese Pharmaceuticals Factory of Shanghai) at a dose of 3 packages (mixed with water and administered orally) per day. In the control group, 27 patients were given the placebo preparation (caramel and white dextrin in a ratio of 1:10; provided by the same factory) which had the same packaging and weight as the SMS preparation. Both the control and treatment group were given a single course of treatment, lasting for four weeks. Prior to, on the third week, and fourth week of the drug treatment, the activities of superoxide dismutase (SOD) and glutathione peroxidase (GSH-Px), and the level of malondialdehyde (MDA) in patients' blood were measured. Using a Doppler device for assessing heart function, heart functional parameters such as stroke volume (SV), cardiac output (CO), and cardiac index (CI) were measured before and after the drug treatment (Table 3.1). Patients in both the placebo control and the SMS-treated groups were given myocardial nutrients and other symptom-oriented medication as usual, but avoided using drugs that could affect the lipid peroxidation. Thirty healthy people were also studied for

*Table 3.2* Activities of blood superoxide dismutase and glutathione peroxidase and level of MDA in patients with viral myocarditis and healthy individuals

| | *Superoxide dismutase (SOD, U/g/Hb)* | *Glutathione peroxidase (GSH-Px, U/g/Hb)* | *Malondialdehyde (MDA, nmol/ml)* |
|---|---|---|---|
| Viral myocarditis (n = 62) | 55.94 ± 16.65* | 11.69 ± 5.78* | 9.40 ± 3.08* |
| Healthy people (n = 30) | 71.94 ± 12.45 | 18.56 ± 2.59 | 5.33 ± 0.97 |

Values given are mean ± SD, with the number of subjects indicated in parentheses.

* significantly different from the healthy group ($p < 0.01$)

*Table 3.3* Effect of SMS treatment on blood superoxide dismutase, glutathione peroxidase activities and malondialdehyde level

| | | *SOD (U/g/Hb)* | *GSH-Px (U/g/Hb)* | *MDA (nmol/ml)* |
|---|---|---|---|---|
| SMS | Before (n = 35) | 54.61 ± 16.56 | 11.28 ± 26.09 | 9.41 ± 2.97 |
| | 2 weeks (n = 18) | 63.15 ± 14.90 | 15.02 ± 8.06 | 8.24 ± 2.36* |
| | 4 weeks (n = 35) | 70.59 ± 18.07* | 16.79 ± 6.41* | 8.30 ± 2.53* |
| CONTROL | Before (n = 27) | 57.59 ± 16.91 | 12.20 ± 5.21 | 9.29 ± 3.26 |
| | 2 weeks (n = 16) | 56.37 ± 13.58 | 13.39 ± 4.59 | 8.08 ± 3.05 |
| | 4 weeks (n = 27) | 63.29 ± 20.37 | 14.40 ± 6.40 | 8.89 ± 1.95 |

Prior to, and on the third and fourth week of SMS treatment, the activities of superoxide dismutase (SOD) and glutathione peroxidase (GSH-Px), and the level of malondialdehyde (MDA) in patients' blood were measured. Values given are mean ± SD, with the number of subjects indicated in parentheses.

* significantly different from that prior to the SMS treatment ($p < 0.01$)

comparison. The results indicated that patients with viral myocarditis were found to have significantly lower blood GSH-Px and SOD activities than those of healthy people ($p < 0.01$), but MDA levels were significantly higher ($p < 0.01$) (Table 3.2). In the SMS treatment group, blood GSH-Px and SOD activities were significantly increased ($p < 0.01$) while MDA level was significantly decreased ($p < 0.01$) (Table 3.3). In the placebo control group, SOD and GSH-Px activities and MDA level did not show any detectable changes before and after the treatment (Table 3.3). The ensemble of results suggested that the myocardial damage caused by viral myocarditis may be related to the lipid peroxidative reactions, in that SMS can scavenge free radicals and inhibit lipid peroxidation, thereby preventing the myocardial damage.

Our hospital has also participated in a clinical project, entitled 'Treatment and Diagnosis of Viral Myocarditis and Dilated Cardiomyopathy', which was a strategic topic in the 'National Nine-Five Research Project' started in 1996. It was a cross-institutional project led by The Shanghai Medical University, Shanghai Second Medical University and The Nanjing Medical University. In this study, patients suffering from viral myocarditis (52 cases) were assigned to the SMS treatment group and treated with a regimen comprising *Radix Astragali*, SMS, and *Calculi Bovis acid*. First of all, 40 ml of SMS injection liquid (every 10 ml was comprised of *Radix Ginseng Destillata* (1 g), *Radix Ophiopogonis* (3 g) and *Fructus Schisandrae* (1.5 g), manufactured by The First Chinese Pharmaceuticals Factory of Shanghai) was supplemented to a 5 per cent glucose solution (250 ml), and was given to the patient by intravenous infusion once

*Table 3.4* Comparison of changes in clinical symptoms and EKG/Holter results between the SMS treatment group and control group

| | | *Chest distress* | *Palpitation* | *Chest pain* | *EKG/Holter* |
|---|---|---|---|---|---|
| | | | *(Number of cases)* | | |
| SMS treatment (n = 52) | Before | 49 | 45 | 23 | 52 |
| | Effective | 34 | 30 | 14 | 19 |
| | Dominant effect | 4 | 7 | 4 | 20 |
| | Ineffective | 11 | 8 | 5 | 13 |
| | Total number of effective cases | 38 | 37 | 18 | 39 |
| | Total effective rate | 78% | 87% | 78% | 75% |
| Control (n = 50) | Before | 50 | 42 | 21 | 50 |
| | Effective | 22 | 25 | 7 | 13 |
| | Dominant effect | 2 | 1 | 3 | 34 |
| | Ineffective | 26 | 16 | 11 | 23 |
| | Total number of effective cases | 24 | 26 | 10 | 27 |
| | Total effective rate | 48% | 62% | 48% | 54% |
| | | $p < 0.01$ | $p < 0.05$ | $p < 0.05$ | $p < 0.05$ |

a day. *Radix Astragali* (20–40 g) was added to a 5 per cent glucose solution and was also given once a day by intravenous infusion. Both drug treatments were given for 14 days. Then one package of SMS oral preparation (manufactured by The First Chinese Pharmaceuticals Factory of Shanghai) was given three times a day. 'Healthy-Heart' granule preparation (manufactured by Zhongshan Hospital of Shanghai, 5 g per package, equivalent to *Radix Astragali* (15 g) and *Radix Sophorae Flavescentis* (15 g)) was given orally, twice a day. *Calculi Bovis* acid was also given orally at a dose of 2 g, three times per day, consecutively for three months. The conditions of the patients were monitored for half a month or every month; clinical symptoms and signs, blood biochemistry and electrocardiogram, and if necessary, a Holter examination, chest X-ray and echo-cardiogram would be carried out. The control group (50 cases) was given myocardial nutrients: Co-A (200U) and ATP (40 mg) added to 5 per cent glucose solution (250 ml) for intravenous infusion; two weeks later, Co-$Q_{10}$ was given orally at a dose of 20 mg, three times per day. Both the SMS treatment and control group were given symptom-orientated drug treatment, and all data were analyzed by t-test to detect the inter-group differences. Results from this study indicated that after one month of treatment the effective rate on chest distress in the SMS treatment group was significantly higher than that of the control group, with the group difference being statistically significant ($p < 0.01$). Improvements in symptoms such as palpitation, chest distress and EKG/Holter results in the SMS treatment group were also significantly different from those of the control group ($p < 0.05$) (Table 3.4). The effective rate on serum cTnT and cTnI in the SMS treatment group was 72 per cent and 59 per cent respectively, while that in the control group was 43 per cent and 32 per cent respectively, with the group differences being $p < 0.05$ (Table 3.5). The ensemble of results from this clinical study indicated that SMS injection liquid and SMS oral preparation could produce good therapeutic efficacy on the treatment of acute viral myocarditis.

*Table 3.5* Comparison of neutralized antibodies, CTnT and CTnI between the SMS treatment group and control group

| | | *Neutralized antibodies* | *CTnT* | *CTnI* |
|---|---|---|---|---|
| SMS treatment (n = 52) | No. of abnormal cases before the treatment | 21 | 29 | 32 |
| | No. of effective cases | 14 | 21 | 19 |
| | Effective rate | 67% | 72% | 59% |
| Control (n = 50) | No. of abnormal cases before the treatment | 26 | 23 | 28 |
| | No. of effective cases | 11 | 10 | 9 |
| | Effective rate | 42% | 43% | 32% |
| | | $p > 0.05$ | $p < 0.05$ | $p < 0.05$ |

## IV. Other reports on the efficacy of SMS treatment

As reported by Huang *et al.* (1999), when given 'Erzhi pills' (*Fructus Ligustri Lucidum* and *Fructus Mori Albae*) and supplemented SMS in company with Co-$Q_{10}$ injection, patients with viral myocarditis were found to have a total effective rate of 91.5 per cent, which was significantly different from the control group ($p < 0.01$). Other studies have reported that various supplemented SMS preparations were used for treating viral myocarditis and the total effective rate ranged from 81.0 per cent to 93.3 per cent (Bu *et al.* 1999; Li 1995; Ying 1999; Zhu *et al.* 1995). As reported by Zhang H. (1997), SMS and *Radix Astragali* injection liquid were used for treating viral myocarditis. The SMS-treated group had an effective rate of 96.8 per cent (Zhang H. 1997). As reported by Yin and Lu (1997), the total effective rate of SMS injection liquid ranged from 92.5 per cent to 96.0 per cent (Peng *et al.* 1995; Tan *et al.* 1995; Yang 1997; Yin and Lu 1997), whereas the use of SMS together with *Lonicerae-Forsythiae* powder could produce an effective rate of 97.3 per cent, as described in Sun (1998). Zhang P.Y. (1997) has reported that *Qingkailing* together with SMS injection could produce an effective rate of 92 per cent. It has also been reported that by adopting the *heart-clearing* and detoxification method for expelling the toxins and SMS injection liquid for treating viral myocarditis, the effective rate was found to be 96.8 per cent. Although these studies have reported different effective rates, all indicated the effectiveness afforded by SMS treatment when compared with the control. The patients given SMS oral preparation or SMS injection liquid could be relieved from clinical symptoms and showed better recovery in heart function, electrocardiogram marker, and biochemical indices. The effective rates of SMS treatment obtained from these studies were generally higher than that reported by our studies (Table 3.1–3.5), which may be due to the difference in observation period for clinical efficacy. Results from all of these studies supported the idea that SMS oral preparation or SMS injection liquid can produce significant therapeutic effect on treating viral myocarditis. SMS treatment can therefore provide a therapeutic means of TCM characteristic for viral myocarditis.

## V. Mechanism of SMS treatment on viral myocarditis

Clinical studies from our laboratory have demonstrated that the protection afforded by SMS treatment against viral myocarditis-induced myocardial damage may be attributed to its inhibitory effect on free radicals and lipid peroxidation.

In the view of the theory of TCM, the treatment for viral myocarditis should be primarily focused on supporting the *healthy energy*, which is supplemented by sedatives and tranquilizing agents, as well as agents that can eliminate the invading pathogens. The invigoration of healthy energy (*Qi*) can modulate the visceral function and reinforce the immune capacity. SMS can be prescribed for benefiting the *vital energy* and recuperating the normal pulse condition. *Ginseng*, one of the component herbs, can invigorate the *primordial-Qi* and *vital energy*, and produce a tranquilizing effect. It can directly activate the myocardium to increase cardiac output and to improve the microcirculation. Recent studies have demonstrated that under conditions of myocardial ischemia-reperfusion, *Ginseng* could enhance the creatine phosphokinase (CPK) and prostacyclin ($PGI_2$) levels, but suppress the thromboxane $A_2$ ($TXA_2$) level, thereby maintaining the beneficial balance between $PGI_2$ and $TXA_2$ (Chen *et al.* 1989; Xu and Liu 1991; Yao *et al.* 1988). In addition, the *Ginseng*-derived saponins could enhance the $PGI_2$ synthesis, but inhibit the thromboxane synthesis (Deng and Luo 1992). The saponins could also inhibit the free radical production from activated polymorpholeukocytes and neutrophils (Deng and Luo 1992). Furthermore, *Ginseng* could increase the level of endogenous glucocorticoid and activate the reticuloendothelial system. *Fructus Schisandrae*, another component herb, can produce an astringent action on the *lungs*, nourish the *kidney*, promote *body fluid* production, prevent excessive perspiration and produce a stimulatory effect on the central nervous system and respiratory system (Yang and Xie 1998). It can also act synergistically with *Ginseng* to activate the adrenocortical functions. *Radix Ophiopogonis*, also a component herb, can produce bacteriostatic effect and hyperglycemic effect, and resist the ischemic damage. Some studies have reported that SMS injection liquid did not produce bacteriostatic effect (Chu *et al.* 1984); instead, it could inhibit the histamine or $^{60}CO$-exposure induced enhancement in capillary permeability, indicating the ability of SMS to produce non-specific anti-inflammatory actions. Glucocorticoid is one of the important endogenous anti-inflammatory agents, and prostaglandin E is an active agent for vascular endothelium relaxation, sensation of pain and inflammation. While SMS injection liquid can significantly increase the level of endogenous glucocorticoid in experimental animals and healthy individuals, this action is particularly important under the condition of adrenocortical insufficiency caused by infectious stress stimulation. On the other hand, SMS injection liquid can reduce the level of plasma prostaglandin E in experimental animals, which is important for the maintenance of tension of vascular endothelial muscle and the suppression of inflammatory responses. These two pharmacological actions produced by SMS injection liquid form the pharmacological basis of therapeutic actions in the treatment of septic shock. The immune system is an important component of the endogenous defensive system against infection, the decline in immune function can increase the chance of infection. While infection can induce specific immune responses, it can also simultaneously inhibit the non-specific immune responses. Research studies have indicated that SMS injection liquid could significantly activate the phagocytic function of the macrophage system and suppress the IgE-mediated humoral immunity, resulting in the priming of immune function to a relatively activated status.

In summary, as judged both on the basis of TCM and results obtained from experimental and clinical studies, SMS oral preparation/SMS injection liquid was found to produce anti-inflammatory and immunomodulatory. It also protects myocardial tissue against ischemic damage, thereby producing beneficial effect on the prevention of disease progression and treatment of viral myocarditis. In addition, the immunomodulatory effect produced by SMS treatment can prevent the occurrence of recurrent viral myocarditis. Experimental and clinical findings from our laboratory suggested that in addition to the use of western medications like anti-infection drugs and myocardial nutrients, SMS injection liquid could be given during the acute phase of viral myocarditis, and SMS oral preparation could be administered later on. The combined treatment could enhance the recovery, reduce the course of disease, and strengthen the health condition of patients suffering viral myocarditis. SMS is therefore regarded as a good example of an old drug for new use.

## HEART FAILURE

Heart failure is a commonly occurring clinical syndrome with a high mortality rate. The incidence of heart failure increases with age, and appears to be higher in males. Pathological changes associated with heart failure, including valvular disease, hypertension, atherosclerosis-related ischemic heart disease, myocardial infarction, cardiac hypertrophy, and pulmonary heart diseases, can eventually lead to the development of heart failure. Clinical manifestations of heart failure include symptoms of insufficient pulmonary circulation such as dyspnea, coughing, expectoration and hemoptysis, as well as symptoms of reduced cardiac output such as hypodynamia, cyanosis, tachycardia, and hypotension. Clinical signs and symptoms of insufficient systemic circulation such as sustained insufficient blood perfusion of organs, edema, jugular filling, swelling and tenderness of the liver, and pleuroperitoneal edema could also be found.

Digitalis, which has been used clinically for two hundred years, is still an effective drug available for the treatment of heart failure nowadays. Digitalis preparations can enhance the cardiac efficiency under conditions of heart failure through increasing myocardial contractility, reducing and prolonging the systolic and diastolic period, respectively, reducing the heart volume, inducing peripheral vasodilatation, increasing the cardiac output, and reducing the myocardial oxygen consumption. In addition to digitalis, diuretics, which are able to decrease the water and sodium retention in the body and thereby reduce the cardiac workload, play an important role in the treatment of heart failure. However, it is found that standard treatment using digitalis and diuretics sometimes cannot manage refractory heart failure. Under such conditions, only multiple drug treatment can work efficaciously. In this regard, SMS injection liquid has been effectively used for the treatment of heart failure in clinical situations. The clinical efficacy of SMS injection liquid was corroborated by experimental studies which suggested the mechanism involved in the cardio-enhancing effect of SMS.

A huge volume of clinical evidence has shown that SMS injection could produce good therapeutic efficacy on heart failure and cardiogenic shock caused by coronary heart disease, myocardial infarction, myocarditis, dilated cardiomyopathy, and pulmonary heart diseases (Wang *et al.* 1994). Prior to the drug treatment, the patients selected for the clinical investigation were assessed for blood viscosity, platelet adhesion and aggregation, and heart function including stroke volume, cardiac output, cardiac index, ejection index, peripheral vascular resistance, left ventricular systolic and diastolic internal radii, and shortening rate of left ventricular axis. In addition, three of the main standard cardiac parameters such as cardiac rate, electrocardiogram, and thoracic X-ray were also examined. In addition to the standard medication, the treatment group was given SMS injection liquid (80 ml) in a 5 per cent glucose adjuvant (500 ml) by intravenous infusion, consecutively for 2 weeks. After the SMS treatment, the aforementioned measurements were repeated for comparison. The results indicated that there were significant ($P < 0.01$) differences in terms of blood viscosity, blood platelet adhesion and aggregation, as well as heart function in patients before and after the SMS treatment, with the mortality rate being also significantly reduced (Wang *et al.* 1994). It was found that the whole blood viscosity, plasma viscosity, and hematocrit were significantly increased in patients with heart dysfunctions. Results obtained from experimental and clinical studies suggested that the reduction of blood viscosity by SMS treatment in patients with heart dysfunctions may be related to a number of actions. SMS treatment can suppress platelet aggregation and release, promote blood circulation by activating the blood flow, and increase both the cardiac output and the velocity of blood flow by enhancing the myocardial contractility. The resultant improvement in anoxic condition could enhance the morphological changes of erythrocytes, which, in turn, reduce the blood viscosity and improve the heart function. After SMS treatment, the majority of the patients were found to have a relief in symptoms such as coughing, expectoration, palpitation, short breath, and dyspnea, to a varying extent. In addition, paroxysmal dyspnea occurring at night was relieved, with the cardiac rate being reduced, the noise from the lungs being decreased/disappeared and the heart sound being strong. Urinary volume of edematous patients was increased, with the extent of edema being decreased to varying degrees. After the SMS treatment, parameters from regular blood and urine analysis, measurements of blood platelet activities, and hepatic and renal functions remained normal. There were no derangements in blood sugar level, blood lipids level, and electrolyte balance. Except for a few individual cases of fever and skin eruption, no severe side effect was observed (Wang *et al.* 1994). The hemodynamic effects produced by SMS in patients suffering from heart failure are described as follows.

## I. Enhancement of heart pumping function

After the intravenous infusion of SMS injection liquid, both cardiac output and stroke volume were increased to varying degrees. Noting that SMS could enhance the stroke volume and concomitantly decrease the peripheral vascular resistance in patients with abnormal left ventricular filling pressure, the pharmacological action produced by SMS seemed to be different from that of vasodilators. Vasodilators can only increase the stroke volume under conditions of higher left ventricular filling pressure than the normal upper limit, flat heart functional curve, and reduced after-load. In contrast to vasodilators, SMS can enhance the cardiac output under different cardiac loading conditions, and improve the pump function under the condition of heart failure. On one hand, these effects could be attributed to the ability of SMS to decrease the resistance against both systemic circulation and left ventricular ejection, so as to increase the stroke volume. On the other hand, the observation that SMS could increase the stroke volume under the condition of low left ventricular filling pressure suggests the ability of SMS to enhance myocardial contractility. The positive ionotropic action of SMS is similar to that of lanatoside, but, without the toxic effect associated with digitalis. Furthermore, SMS could enhance the prostacyclin synthesis and inhibit the thromboxane synthesis, as well as selectively antagonize the blood platelet-activating factors. As a consequence, the blood viscosity and the extent of platelet adhesion and aggregation are significantly reduced, leading to a condition favorable to the removal of *blood stasis* in small blood vessels and hence reduction in the resistance of micro-circulation and improvement in both myocardial blood flow and metabolism. As SMS injection liquid can improve the heart function through fundamental means, it serves as an effective drug for the treatment of heart failure, with its effective period being longer than that of the conventional drug, dopamine.

## II. Enhancement of heart function without the increase in myocardial oxygen consumption

Clinical studies have shown that one hour after the administration of SMS injection liquid, patients suffering from myocardial infarction were found to have LVET being significantly prolonged, PEP, ICP being reduced, PEP/LVET ratio being decreased, LVET/ICT ratio being increased, and cardiac rate and peripheral vascular resistance being reduced. All these significant changes indicated that SMS injection liquid could enhance the left ventricular contractility, stroke volume and cardiac output, without causing any increase in the myocardial oxygen consumption but maintaining the balance between oxygen demand and supply to the myocardium.

## III. Invigoration of the heart-Qi

According to the theory of TCM, the *heart* controls the blood circulation, and the continuous movement of blood inside the vessels is driven by the *heart-Qi* which can be regarded as the pumping function of the heart. The pathogenesis of heart failure involves both *asthenia* in the body constitution (*visceral*) and sthenia in superficiality (*surface*). Diagnosis based on *asthenia* syndrome indicates that heart failure is likely to be associated with *Qi*-deficiency and deficiency of both the *Qi* and the *Yin*. *Qi* is the driving force for the functionality of organs; *primordial-Qi* deficiency can cause the decline in visceral functions and the insufficiency in energy transformation (hence the *Qi* and the *Yin* deficiency), which, in turn, affect the nourishment of the *heart* (Cao 1997). Deficiency of the *Qi* is found to be associated with an impediment in blood circulation, as clinically manifested by the weak pulse condition and *blood stasis*, as well as palpitation. The *spleen asthenia* can cause the inability to transform the *dampness*, whereas *kidney asthenia* can lead to edema arising from the retention of *body fluid*. The accumulated *dampness* can attack the *lung* and cause the shortness of breath, with coughing and dyspnea on lying posture (Cao 1997).

*Radix Ginseng*, which serves as the '*Monarch*' (the chief drug) in the SMS formula, can supplement the *lung* function and invigorate the *Qi* in the *channels* connecting to the *lung*, *spleen* and *heart*. It is therefore capable of invigorating the *primordial-Qi*, supplementing the *spleen* and *lung* function, promoting production of *body fluid* and tranquilization. *Radix Ophiopogonis*, which serves as the 'Minister' (assistant drug), can act through the *channels* to the *lung*, *heart* and *kidney*, so as to benefit the *Qi* and promote the production of *body fluid*, supplement the *kidney* function, and nourish the *heart*. *Fructus Schisandrae* can produce the astringent function, retain the *lung-Qi* and stop excessive perspiration for promoting production of *body fluid*. The combined use of these three herbs, which produce a supplementing, clearance and astringent action, respectively, can benefit the *Qi*, nourish the *Yin*, promote the production of *body fluid* production to quench thirst and preserve the *Yin* to prevent collapse. These effects can result in the replenishment of the *Qi*, restoration of the normal pulse, and preservation of the *Yin* by reducing perspiration, eventually leading to the restoration of visceral functions and the strengthening of the body constitution. Modern pharmacological studies have indicated that the principal ingredients of SMS injection liquid include ginsenosides, organic acid, schisandrins, various trace elements, and other saponins. These active components of SMS can enhance the organ function and modulate the body metabolism, thereby relieving the symptoms of chest distress, palpitation, hypodynamia and excessive perspiration. These actions are analogous to the replenishment of the *Yin*, the *Yang*, the *Qi*, and the *blood* for relieving various *asthenia* symptoms in TCM. Hence, the fundamental principle underlying the treatment of heart failure should be focused on invigorating the *Qi*, while the enhancement of cardiac pump function and cardiac output are equivalent to the *Qi*-invigorating as described in TCM.

In summary, SMS injection liquid can effectively treat heart failure caused by cardiac hypertrophy, rheumatic and pulmonary heart diseases. It can also be used for treating heart failure patients who are non-responsive to digitalis treatment.

## CARDIOGENIC SHOCK

Shock is a sudden decrease in tissue perfusion and oxygen supply induced by an acute circulatory disturbance, with clinical manifestations of a series of metabolic and functional disorders. While the incidence of shock can be originated from various diseases, cardiogenic shock is more prevalent. Despite the recent advance in the therapy and prevention of shock, the mortality rate of shock still remains at a high level. Hence, further investigation on the treatment of shock is necessary.

According to our clinical experience and reports from other hospitals in China, SMS has been widely used as a supplementary regimen for the treatment of shock (Li 1997; Wang and Zhou 1995). While being used together with standard therapeutic intervention for shock symptoms, SMS injection liquid could enhance the cardiac output, with blood pressure being progressively increased, limb temperature being restored, urine volume being increased, perspiration being reduced, and mental status being tranquilized. The clinical efficacy afforded by SMS injection liquid was significantly higher than that of the control group (i.e. without supplementary SMS treatment). After the termination of vasoactive drug administration, the blood pressure was maintained to a better extent than that of the control group. In general, early after the onset of myocardial infarction, the peripheral vascular resistance is increased by the endogenously secreted catecholamines. Under such conditions, the use of sympathomimetic drugs for elevating blood pressure can further jeopardize the peripheral circulation, leading to symptoms such as clammy limbs and reduction of urine volume. Through resolving the contradictory action of hypertensive drugs in the increase of both coronary blood flow and peripheral resistance, SMS injection liquid can enhance the cardiac output, with the arterial pressure being increased, and peripheral resistance being decreased. The blood pressure modulatory effect of SMS injection liquid is relatively mild, without producing abrupt changes in blood pressure as in the case of hypertensive drug treatment. Being endowed with the *Qi* invigorating, pulse recuperating and collapse preventing actions, SMS injection liquid is a reliable treatment for cardiogenic shock. However, the relatively slow blood pressure elevating effect of SMS warrants the application of other anti-shock therapeutic interventions, particularly on the primary cause of the disorder.

## RESPIRATORY FUNCTION IN CHRONIC OBSTRUCTIVE PULMONARY DISEASE

Patients suffering from chronic obstructive pulmonary disease (COPD) are often complicated by the decline in lung and respiratory function, which is related to the contractile dysfunction of the diaphragm. SMS has been shown to enhance the contractile strength of the diaphragm in rats (Zhao and Niu 1995). Fang *et al.* (1998) reported that patients suffering from COPD were treated with SMS (100 ml/day, i.v.) for 14 days and parameters, including lung vital capacity (VC), forced vital capacity (FVC), forced vital capacity of the first second ($FEV_1$), $FEV_1$/FVC, maximum voluntary ventilation (MVV), maximum inspiratory pressure (MIP), load respiratory time (LT), 6 minute walk distance (6MWD), arterial blood gas and Borg dyspnea scale, were examined before and after the treatment. It was found that all parameters except for FVC and $FEV_1$/FVC were improved by the SMS treatment, and they were significantly better than the control group. The results suggested that SMS treatment could improve respiratory function in COPD.

## REFERENCES


Bu, C.D., Zhao, Y.J., and Sun, F.J. (1999) The application of modified Shengmai San for the treatment of 13 cases of viral myocarditis, *Chinese Journal of Integrated Traditional and Western Medicine in Intensive and Critical Care*, 6(2), 92.

Cao, G.M. (1997) The traditional Chinese medicine perspective and its therapeutic approach of heart failure, *Journal of Shanxi College of Traditional Chinese Medicine*, 20(3), 6.

Chen, B. and Chen, Y.F. (1996) Treatment of 2 cases of heart failure associated with anemic cardiomyopathy, *Xiandai Zhongyi*, 30(1), 48–50.

Chen, G.Z. and Wu, X.R. (1996) Treatment of refractory ventricular arrhythmia with the combined use of FDP, berberine, XinBaoWan and modified SMS, *Journal of Shantou University Medical College*, 2, 90–1.

Chen, L., Rong, Y.Z., and Han, Y.S. (1989) Cardioprotective effect of diltiazem against ischemia/reperfusion injury, *Shanghai Journal of Medicine*, 12(12), 697–700.

Chen, S.X., Chang, P.L., Zheng, X.J., *et al.* (1990) Effects of the treatment of Coxsackie virus-induced myocarditis with Yupingfeng San and Shengmai San, *Chinese Journal of Integrated Traditional and Western Medicine*, 10(1), 20–1.

Chen, Y.N. (1997) Clinical study on the combined use of Shengmai Tonic and Astragali Tonic in the treatment of arrhythmia associated with chronic hepatitis, *Liaoning Journal of Traditional Chinese Medicine*, 24(9), 408.

Chu, Y., Pan, D.B., and Yang, Y.Q. (1984) Anti-inflammatory action of Shengmai injection and its effect on the immune system, *Yixue Tongbao*, 19(7), 23–6.

Cong, Y.Z. (1980) Studies on Shengmai injection, *Chinese Patent Medicine Research*, 2, 41–3.

Deng, X.W. and Luo, D.C. (1992) Effects of SMS on the left ventricular functions, exercise tolerance and oxygen-derived free radicals in angina patients, *Chinese Journal of Internal Medicine*, 31(4), 113–6.

Department of Cardiology, Xinhua Hospital (1999) Clinical studies on the treatment of angina pectoris with SMS injection, *Proceedings of the 5th National Conference on Integration of Traditional and Western Medicine for the Prevention and Treatment of Coronary Heart Disease, Angina Pectoris and Arrhythmia*, p.32–5.

Dong, Q.Z., Chen, K.J., Tu, X.H., *et al.* (1984) Effects of Shengmai San treatment on the hemodynamics of patients of acute myocardial infarction, *Chinese Journal of Cardiology*, 12(1), 5–8.

Du, D.S., Song, Y., Xu, Q., and Meng, X. (1998) Clinical effects of SMS injection on arrhythmia associated with hyperthyroidism, *Journal of Practical Integrated Traditional and Western Medicine*, 11(8), 702–3.

Editorial of the Chinese Journal of Internal Medicine (1987) Proceedings of the National Forum on Myocarditis and Myocardial Diseases, *Chinese Journal of Internal Medicine*, 26, 597–601.

Fang, J., Jiang, J., and Luo, D.C. (1987) Effects of SMS tonic in the heart functions of patients with coronary heart disease, *Chinese Journal of Internal Medicine*, 26(7), 403–6.

Fang, Z.J., Jiang, H.M., and Wang, L.H. (1998) Therapeutic effect of Shengmai injection on respiratory function in chronic obstructive pulmonary disease, *Chinese Journal of Integrated Traditional and Western Medicine*, 9, 520–2.

Guo, S.K. (1981) Clinical study on the use of dl-demethylcolarine and SMS infusion in the treatment of bradycardia, *Beijing Medical Journal*, 1, 46–7.

Hua, M.Z. and Qi, H. (1996) Treatment of 65 cases of acute myocardial infarction in senile patients with a modified formulation of Shengmai San, *Chinese Journal of Integrated Traditional and Western Medicine in Intensive and Critical Care*, 3(6), 259–60.

Huang, J.X., Han, Z.X., Su, X.Y., *et al.* (1999) Effects of the combined use of Shengmai San, modified Erzhi Wan and coenzyme-Q10 injection for the treatment of viral myocarditis, *Chinese Journal of Integrated Traditional and Western Medicine in Intensive and Critical Care*, 6(3), 129–31.

Huang, N., Chen, Y., and Zhou, M. (1993) Protection of the myocardium against ischemia-reperfusion injury by Shengmai San, *Journal of the First Military Medical University*, 13(1), 43–6.

Jia, W.H. (1998) Clinical effects of Astragli-supplemented Shengmai San on the treatment viral myocarditis, *Chinese Journal of Experimental Traditional Medical Formulae*, 4(2), 35–7.

Jiang, J., Fang, J., and Luo, D.C. (1988) Effects of SMS on the left ventricular functions and exercise endurance of patients with dilated cardiomyopathy, *Chinese Journal of Cardiology*, 16(2), 65–7.

Li, C.Q. (1997) Clinical observations of the treatment of 32 cases of shock associated with early phase of infection, *Dangdai Yishi Zazhi*, 2(6), 50.

Li, G.H., Xuan, M.D., Zhou, H.C., *et al.* (1994) Protection of the myocardium by Shengmai tonic in open-heart surgery, *Journal of Hunan Medical University*, 19(5), 413–6.

Li, H.L. (1978) Treatment of cardiogenic shock associated with acute myocardial infarction, *Heilongjiang Medical Journal*, 6(3), 7–8.

Li, M.J., Yang, J.X., Wang, X.L., *et al.* (1999) Effects of SMS injection on the exercise endurance, cardiogram and blood viscocity of patients with angina pectoris, *Journal of Emergency Syndromes in Traditional Chinese Medicine*, 14(2), 138.

Li, S.P. (1995) Treatment of 42 cases of viral myocarditis with a modified formulation of Shengmai San, *Shanxi Zhongyao*, 16(9), 411.

Liu, X.G., Zhao, Q., Zhou, Z.P., *et al.* (1978) Effects of SMS on DNA metabolism in the myocardium, *Shanxi Xin Yiyao*, 4, 6–9.

Lu, B.J., Rong, Y.Z., Zhao, M.H., Huang, G.F., Zhu, X.Y., and Yang, J.W. (1994) Protection against lipid peroxidation by SMS in acute myocardial infarct patients, *Chinese Journal of Integrated Traditional and Western Medicine*, 14(12), 712–4.

Lu, H.X., Zhang, G.L., and Liu, H.Y. (1997) Treatment of heart failure in dilated cardiomyopathy with the combined use of SMS and dopamine, *Chinese Journal of Integrated Traditional and Western Medicine in Intensive and Critical Care*, 4(5), 112–3.

Lu, S.B., Lu, B.J., Li, J.X., *et al.* (1998) Effects of SMS injection on the blood viscosity, platelet aggregation and left ventricular function in patients with coronary heart disease, *Tianjin Yiyao*, 26(4), 226–8.

Mo, Z.Z., Zhuang, R.X., Li, R.J. *et al.* (1997) Combined use of SMS and atherosclerotic plague solubilizing agent – urokinase, in the treatment of acute myocardial infarction, *Chinese Journal of Integrated Traditional and Western Medicine in Intensive and Critical Care*, 4(11), 487–9.

Pathophysiology Research Group, Beijing Medical University (1977) Study of cardiogenic shock IV, *Journal of Beijing Medical University*, 2, 101–4.

Pathophysiology Research Group, Beijing Medical University (1978), Experimental studies on cardiogenic shock VII, *Journal of Beijing Medical University*, 4, 222–5.

Peng, W.H., Zhang, Z.G., Wang, Z.R., *et al.* (1995) Treatment of 40 cases of viral myocarditis with Shengmai injection, *Medical Recapitulation*, 1(10), 463–4.

Qin, L.M., Yang, J.D., and Liao J.Z. (1983) Effects of SMS on the ATPase activities of cardiomyocytes in rats, *Chinese Journal of Critical Care Medicine*, 3(2), 6–8.

Qiu, R.X. and Luo, Z.Q. (1989) Clinical application of SMS for the treatment of coronary heart disease, *New Journal of Traditional Chinese Medicine*, 7, 14–6.

SATCM, P.R.China (1995) Experimental and 219 cases of clinical studies of SMS injection in the treatment of coronary heart disease, *Journal of Emergency Syndromes in Traditional Chinese Medicine*, 4(4), 152–5.

Shi, Z.X., Liao, J.Z., Wu, Z.M., *et al.* (1981) Experimental and clinical studies of Shengmai San, *Journal of Traditional Chinese Medicine*, 12, 947–51.

Su, H.H. (1998) Clinical study of the treatment of asymptomatic myocardial ischemia with Shengmai injection, *Chinese Journal of Integrated Traditional and Western Medicine*, 11(10), 897.

Sun, C.X., Zhang, H.M., and Li, Y.H. (1995) Study on the treatment of 20 cases of arrhythmia with SMS, *Hebei Journal of Traditional Chinese Medicine*, 17(2), 52–6.

Sun, D.X. (1998) The combined use of Shengman San and Yinqiao for the treatment of 37 cases of viral myocarditis, *Forum on Traditional Chinese Medicine*, 13(2), 26.

Sun, L. and Wang, M. (1998) Clinical studies on the treatment of unstable angina with SMS injection, *Journal of Chinese Medicine and Pharmacology*, 6, 23–4.

Sun, Y. (1995) Study on the treatment of viral myocarditis: application of Shengmai San-related formulations, *Zhejiang Journal of Traditional Chinese Medicine*, 30(9), 423–4.

Tan, J.C., Xie, H.F., and Zhang, C.Y. (1995) Treatment of 30 cases of viral myocarditis with Shengmai injection, *Hebei Journal of Traditional Chinese Medicine*, 17(4), 47–8.

Tan, S.J. (1997) Clinical effects of Shengmai injection and dobutamine on the treatment of severe heart failure, *Journal of Practical Integrated Traditional and Western Medicine*, 10(21), 2135–6.

Tianjin Nankai Hospital (1973) Clinical studies in 105 cases of acute myocardial infarction with an integrated approach of traditional Chinese and modern medicine, *Tianjin Yiyao*, 1, 10–18.

Wang, J., Sheng, J., Xu, Z.J., *et al.* (1999) Effects of SMS treatment on the symptoms and the levels of antithrombin III, thromboxane $B_2$ and prostaglandin $F_{1\alpha}$ in patients with coronary heart disease, *Chinese Journal of Integrated Traditional and Western Medicine*, 19(4), 256.

Wang, J.H. and Yu, C. (1997) Effects of SMS injection on the treatment of positive ventricular late potential associated with coronary heart disease and hypertension, *West China Journal of Pharmaceutical Sciences*, 12(3), 201–2.

Wang, W.X., Yan, Q.H., Hu Q.K., *et al.* (1997) Treatment of angina pectoris with SMS injection, *Chinese Journal of New Drugs and Clinical Remedies*, 16(4), 249–50.

Wang, Y.M. and Zhou, L.Y. (1995) Treatment of cardiogenic shock by high doses of Shengmai injection, *Journal of Changchun College of Traditional Chinese Medicine*, 11(49).

Wang, Z.C. (1998) The combined use of Shuanghuanglin injection and Shengmai injection for the treatment of 40 cases of acute viral myocarditis, *Journal of Changchun College of Traditional Chinese Medicine*, 14(1), 13.

Wang, Z.H., Zhang, J.Z., and Tu, Y.S. (1994) Effects of Shengmai injection treatment on the clinical manifestations and hemodynamics of patients with impaired heart functions, *Journal of Clinical Cardiology*, 9(3), 190.

Xiong, D.D., Su, Y.G., Yang, Y.Z., and Chen, H.Z. (1997) Combined use of taurine from Astragali and Co-Q10 for the treatment of viral mycarditis, *Chinese Journal of Cardiology*, 25(5), 329.

Xu, Q.Y., Wang, H.C, and Liu, W.X. (1986) Effects of SMS injection on the formation of blood-stasis in vitro and the blood clotting system in rabbits, *Chinese Journal of Integrated Traditional and Western Medicine*, 6(7), 428–9.

Xu, X.S. and Liu, C.X. (1991) Mechanism of ischemia/reperfusion injury in the myocardium and its prevention and treatment with traditional Chinese medicine, *Chinese Journal of Integrated Traditional and Western Medicine*, 11(2), 124–6.

Yang, F.Q. and Xie, W.H. (1998) Effects of Shengmai injection and the method of heat-evil clearing on the treatment of juvenile viral myocarditis, *Inner Mongolia Journal of Traditional Chinese Medicine*, 17(3), 8–9.

Yang, H.B. (1997) Clinical study on the treatment of viral myocarditis with Shengmai injection, *Journal of Chinese Medicine and Pharmacology*, 25(3), 11.

Yao, M., Rong, Y.Z., Wen, W.H., *et al.* (1988) Prevention of ischemia-reperfusion injury in cardiomyocytes by Shengmai San treatment, *National Medical Journal of China*, 68(6), 313–5.

Yin, Y.C. and Lu, Z.F. (1997) Treatment of viral myocarditis with Shengmai injection, *Journal of Practical Integrated Traditional and Western Medicine*, 10(15), 1477–8.

Ying, A.L. (1999) Treatment of 30 cases of viral myocarditis with a modified formulation of Shengmai San, *Journal of Practical Traditional Chinese Medicine*, 15(3), 8.

Yuan, Y.X. and Zheng, B.D. (1997) Integration of traditional Chinese medicine and western medicine for the treatment of intractable heart failure associated with dilated cardiomyopathy, *Chinese Journal of Integrated Traditional and Western Medicine in Intensive and Critical Care*, 4(5), 202–3.

Zeng, J.S. (1996) Treatment of coronary heart disease with a modified formulation of Shengmai San: effects on the ventricular late potential, *Chinese Journal of Integrated Traditional and Western Medicine in Intensive and Critical Care*, 3(10), 448–9.

Zhang, D.N., Ni, J.X., Zhang, L.R., *et al.* (1984) Treatment of myocardial infarction with Shengmai injection, *Journal of Sichuan Medical University*, 15(2), 131–5.

Zhang, G.Z., Zhu, K.R., and Yan, P.Q. (1997) Effects of Shengmai injection on 40 cases of dilated cardiomyopathy, *Clinical Focus*, 12(7), 316–8.

Zhang, H. (1997) The application of 'Treatment by the differentiation of signs and symptoms' in different stages of dilated cardiomyopathy, *Anhui Zhongyi Linchuang Zazhi*, 9(3), 135.

Zhang, J., Wang, Z.H., and Yu, S.M. (1990) Effects of SMS on the hemodynamics of patients with dilated cardiomyopathy-associated heart failure, *Journal of Hunan Medical University*, 15(2), 153–5.

Zhang, P.Y. (1997) Treatment of acute viral myocarditis with the combined use of Qingkailing and Shengmai injection, *Journal of Emergency Syndromes in Traditional Chinese Medicine*, 6(6), 265–6.

Zhang, X. (1998) Study on the use of Astragali injection for the treatment of viral myocarditis, *Chongqing Medical Journal*, 27(2), 124–5.

Zhang, Z.Y. and Yang, Z.W. (1986) Effect of SMS injection in the hemodynamics and oxygen consumption of the cardiomyocytes in anesthetized animals, *West China Journal of Pharmaceutical Sciences*, 1(2), 86–90.

Zhao, J.P. and Niu, R.J (1995) Effects of Shengmai injection on the contractility of the diaphragm under chronic hypoxic conditions in rats, *Chinese Journal of Tuberculosis and Respiratory Diseases*, 18(1), 53.

Zhao, M.H, Rong, Y.Z., Lu, B.J., *et al.* (1996) Effects of Shengmai San on the serum lipid peroxides levels in patients with acute myocarditis, *Chinese Journal of Integrated Traditional and Western Medicine*, 16(3), 142–5.

Zhou, H.Y. (1996) Treatment of frequent premature beat with the combined use of Astragali-supplemented SMS, amiodarone and high doses of oryzanol, *Chinese Journal of Integrated Traditional and Western Medicine in Intensive and Critical Care*, 3(7), 296.

Zhou, Y., Deng, S.Y., Shu, S., *et al.* (1997) Effect of SMS on the dilating function of the left ventricle, *Gui Zhou Yi Yao*, 21(2), 83–5.

Zhu, Q.L. and Qiao, L.X. (1998) Combined use of Shengmai San and fructose-1,-6-bisphosphate in the treatment of acute myocardial infarction with atrioventricular blockade, *Chinese Journal of Integrated Traditional and Western Medicine in Intensive and Critical Care*, 5(3), 108.

Zhu, Q.R., Wang, Q., and Wan, Y.E. (1997) Combined use of SMS and western medicine in the treatment of arrhythmia, *Chinese Journal of Integrated Traditional and Western Medicine in Intensive and Critical Care*, 4(5), 233–5.

Zhu, Q.R., Wang, Y.E., and Wang Q. (1995) Effects of the combined use of SMS and nicotinamide on the treatment of senile bradycardia in 168 patients, *Chinese Journal of Integrated Traditional and Western Medicine in Intensive and Critical Care*, 2(6), 246–8.

Zhu, R.F., Liu, Z.J., and Sun, Y. (1995) Treatment of 63 cases of viral myocarditis with a modified formulation of Shengmai San, *Shandong Journal of Traditional Chinese Medicine*, 14(5), 207–8.

# 4 Side effects of Shengmai San

*Zhi-Min Xu and Ye-Zhi Rong*
[Translated by Kam-Ming Ko]

Shengmai San (SMS), a traditional Chinese medicine (TCM) formula now existing in various pharmaceutical preparations, is comprised of *Radix Ginseng*, *Radix Ophiopogonis* and *Fructus Schisandrae* (1:3:1.5, w/w). *Radix Salviae Miltiorrhizae* is usually selected as a substitute for the *Ginseng* component in SMS because of its similar *Qi*-invigorating property. *Radix Ginseng/Radix Salviae Miltiorrhizae* is the principal pharmacological component in the formulation of SMS that can produce positive inotropic, peripheral vasodilating, and blood viscosity-lowering effects. However, because of its *warm-heat nature* and bi-directional pharmacological properties, administration of *Radix Ginseng/ Radix Salviae Miltiorrhizae* may exaggerate the symptoms of *Yang hyperactivity* in the already *Yang-hyperactive* patients. On the other hand, *Radix Ophiopogonis* and *Fructus Schisandrae* are effective *Yin*-nourishing herbs, the combined use of these herbs with *Radix Ginseng/Radix Salviae Miltiorrhizae* can offset the side effects of *Radix Ginseng/ Radix Salviae Miltiorrhizae*, thereby preventing the incidence of imbalance between *Yin* and *Yang* in the patients given SMS treatment. Nevertheless, patients with apparent hyperactive *Yang* should avoid using SMS to prevent the exaggeration of *Yin-Yang* imbalance. *Yang hyperactivity* is clinically manifested as distention of head, headache, dizziness and restlessness, and is usually associated with strong pulses, tongue redness with little fur and yellowish urine. Furthermore, SMS should not be prescribed for patients with pyretic and sthenia syndromes, such as uncontrolled acute inflammatory responses (fever, infection, icterus, etc.). Hence, SMS treatment should only be applied on the basis of *Differentiation of Signs and Symptoms* so as to avoid the incidence of unfavorable side effects.

Clinical observations on the long-term use of SMS have indicated that SMS did not produce any unfavorable effects on the hepatic/renal functions, hemopoietic system or endocrine system. However, some scarce incidences of unfavorable side effects have been observed in the clinical use of SMS-related preparations. In the present review, we classified the side effects into two main categories: 1) extrinsic, those related to the nature of the pharmaceutical preparation *per se*, and 2) intrinsic, unrelated to the pharmaceutical preparation. We also tried to define the underlying mechanisms of these side effects from the TCM perspective. It was found that most of the side effects observed under clinical situations were related extrinsically to the nature of pharmaceutical preparation rather than intrinsically to SMS itself.

## EXTRINSIC SIDE EFFECTS

### I. Anaphylaxis and drug rash

Almost all drug-induced anaphylactic responses/rash associated with injection of pharmaceutical preparations were observed during or after the injection process. These side effects include scarlet fever-like rash or multiple typhus (Hao *et al.* 1997), which are characterized by their irregular and crowded distribution, bright red color (colorless upon pressing), and usually, itchy sensation. Rash usually appears around the trunk, neck, face, and upper limbs, possibly associated with light to mild fever (<38.5°C), and occurring within one to two weeks after the drug treatment or without any apparent latent period. Under normal circumstances, the rash would disappear 3 to 4 days after discontinuation of drug treatment, without leaving any scar on the skin. Yet, in case of generalized severe rash, hormonal or anti-allergic treatment would be needed. Xu (1998) has reported a case of exfoliative dermatitis associated with the use of SMS injection. Medical personnel should be aware of the importance of such reports. A minority of patients may develop the dangerous side effect of anaphylactic shock (Ren and Xia 1997; Xin 1999). Because of their life-threatening potential, prompt medical attention should be taken should symptoms of anaphylactic shock appear during SMS infusion. Such symptoms include idiopathic chest distress, shortness of breath, pale complexion or blood pressure decline.

The development of drug rash may be related to the action of individual constituents or attributed to the presence of impurities in the injection preparations. Due to the complicated and diverse chemical composition of Chinese herbs, non-optimized manufacturing processes and differences in quality standards among manufacturers, the presence of micro-particles represents a common problem in TCM-derived injection preparations.

In China only a small number of injection preparations, like *Danseng* Injection, are free from the problems of micro-particles. SMS injection is the fine-processed extracts of *Radix Ginseng Destillata*, *Radix Ophiopogonis* and *Fructus Schisandrae*, with a complex array of active ingredients, and possibly contains some macro-biomolecules like protein and polysaccharides. These varied sized micro-particles may act as hapten that can conjugate with plasma proteins, induce allergic reactions, and hence contribute to the development of anaphylaxis and drug-induced rash. Clinical observations also showed that SMS injections from different manufacturers, or different batches from the same manufacturer, may cause different rates of anaphylactic response. Therefore, upgrading the manufacturing process and the quality standards of SMS injection preparations are the key factors in minimizing the occurrence of side effects (Liu *et al.* 1999).

## INTRINSIC SIDE EFFECTS

### I. Dizziness, headache and insomnia

During the course of SMS treatment, a few patients have reported to have mild symptoms of headache, dizziness, distention of the head and insomnia. While the symptoms usually disappeared after discontinuation of use, the side effects were not related to the dosage form being administered. Instead, they may be attributed to the stimulation of the central nervous system by ginseng saponins.

## II. Abdominal distension, and lumbar and back myalgia

Patients given high doses of SMS may have abdominal distension and anorexia (Zhou and Yang 1996). According to the TCM theory of *Treatment by Differentiation of Signs and Symptoms*, *Ginseng* is a potent *Qi*-invigorating herb, and its use at high doses may cause over-invigoration of the *Qi*. Such over-invigoration would cause an overflow of the *liver-Qi* to the *stomach* and an elevation of the *stomach-Qi*, manifested as abdominal distension. The symptom reflects a temporary state of imbalance during the course of treatment, and is usually relieved upon discontinuation of drug use. Furthermore, the cause of abdominal distension may also be attributed to the *cold nature* of *Radix Ophiopogonis* and the acidic nature of *Fructus Schisandrae*. *Fructus Schisandrae* contains a large amount of acidic components such as malic acid, citric acid, tartaric acid and ascorbic acid, which may cause gastric irritation. Chen (1999) reported that some patients developed the side effects of lumbar and back myalgia, of which the cause cannot readily be explained by modern medicine from the pathological and physiological standpoints. However, the side effects may be explained by the theory of TCM in that the over-invigoration of the *Qi* would cause the excessive accumulation of *Qi* inside the body, resulting in muscular spasm. No particular treatment is required, and the side effects tend to be alleviated upon discontinuation of SMS treatment or a reduction in dosage.

## III. Blood pressure elevating effects on some hypertensive patients

Clinically, SMS treatment was found to cause an elevation in the blood pressure, particularly the systolic pressure, in a small number of hypertensive patients. Given that *Radix Ginseng Destillata* is able to increase myocardial contractility and the stroke volume, failure to attenuate these effects by peripheral vasodilation may partly explain the effect of elevated blood pressure by SMS. Furthermore, the astringent/antihidrotic effects of *Fructus Schisandrae* and the *Yin*-nourishing effect of *Radix Ophiopogonis*, which can maintain the normal vascular permeability of blood vessel and increase the effective blood volume, may contribute to the elevation of blood pressure. These pharmacological effects produced by *Fructus Schisandrae* and *Radix Ophiopogonis* are, in fact, ascribable to one of the action mechanisms of SMS in treating circulatory shock. However, such a blood pressure-elevating effect is unfavorable for patients with severe hypertension. Hence, SMS treatment should be accompanied by antihypertensive treatment when used in hypertensive patients, and should be used with caution in cases of severe hypertension.

## REFERENCES

Chen, Y.P. (1999) Report on 2 cases of back myalgia associated with SMS injection, *Chinese Journal of Hospital Pharmacy*, 19(2), 127.

Hao, J., Wang, Z.A., and Dan, S.J. (1997) Report on 2 cases of allergic skin reaction associated with the use SMS injection, *Chinese Journal of Hospital Pharmacy*, 17(4), 188.

Liu, Y.L., Teng, L., Li, C.J., and Yu, Y.T. (1999) Unfavorable effects of Shengmai San (SMS) injection, *Yi Xue Dao Bao*, 18(2), 131.

Ren, X.H. and Xia, Q. (1997) Report on a case of hypersensitivity associated with the intravenous infusion of Shengmai San injection, *Journal of North Sichuan Medical College*, 12(1), 99.

Xin, F.B. (1999) A case of hypersensitivity associated with the use of SMS injection, *Traditional Chinese Drug Research and Clinical Pharmacology*, 10(2), 115.

Xu, F.N. (1998) Report on a case of exfoliative dermatitis induced by the use SMS injection, *Journal of Anhui College of Traditional Chinese Medicine*, 17(S:S), 95.

Zhou, X.L. and Yang, Y.Q. (1996) Report on 4 cases of severe stomach distention associated with high doses of SMS injection, *Forum on Traditional Chinese Medicine*, 11(2), 36.

# 5 Phytochemistry and pharmacology of component herbs of Shengmai San

*Siu-Po Ip, Timothy C.M. Tam and Chun-Tao Che*

## RADIX GINSENG

*Radix Ginseng* (*Renshen*) is the root of *Panax Ginseng* C. A. Mey. The root turns reddish after steam treatment, and is known as 'red ginseng'. It is sweet and slightly bitter in *taste.* The use of *Radix Ginseng* has a long history in traditional Chinese medicine. According to *Shen Nong Ben Cao Jing* published around 101 B. C. during the *West Han* Dynasty, *Ginseng* was said to be able to 'support the five visceral organs, calm the nerves, tranquilize the mind, stop convulsions, expunge evil spirits, clear the eyes, and improve the memory' (Huang 1998a). The efficacy of *Ginseng* was known in the West by the $18^{th}$ century, and the pharmacological effects of *Ginseng* have been demonstrated in the central nervous system (CNS) as well as in cardiovascular, endocrine, and immune systems.

### I. Chemical constituents of *Radix Ginseng*

The chemical constituents, especially saponins of *Ginseng* and the congeners, have been investigated extensively. They can be classified into the following major groups.

#### *A. Saponins*

The major constituents of *Ginseng* are the saponins. There are three structural types of sapogenins: oleanolic acid, 20(S)-protopanaxdiol and 20(S)-protopanaxtriol.

(1) Oleanolic acid type: ginsenoside $R_o$ (Figure 5.1) (Sanada 1974, Li 1979)

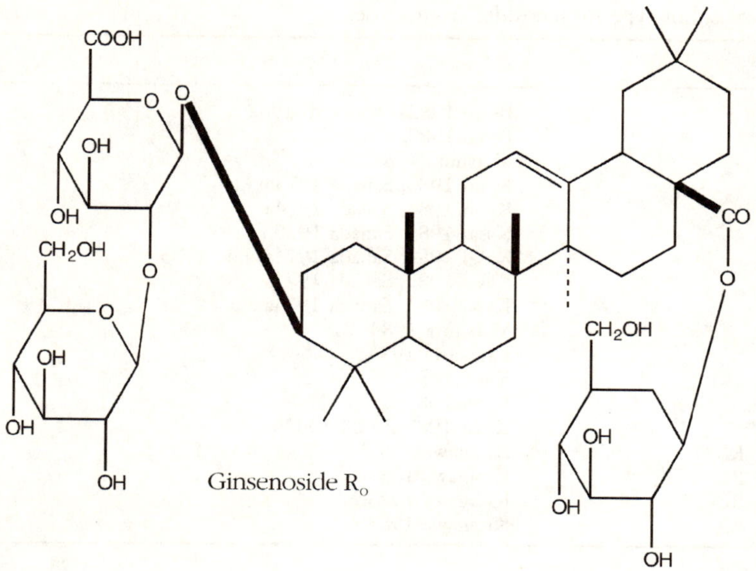


*Figure 5.1* Chemical structure of ginsenoside $R_0$, an oleanolic acid type of saponin found in *Radix Ginseng*.

(2) Protopanaxdiol type

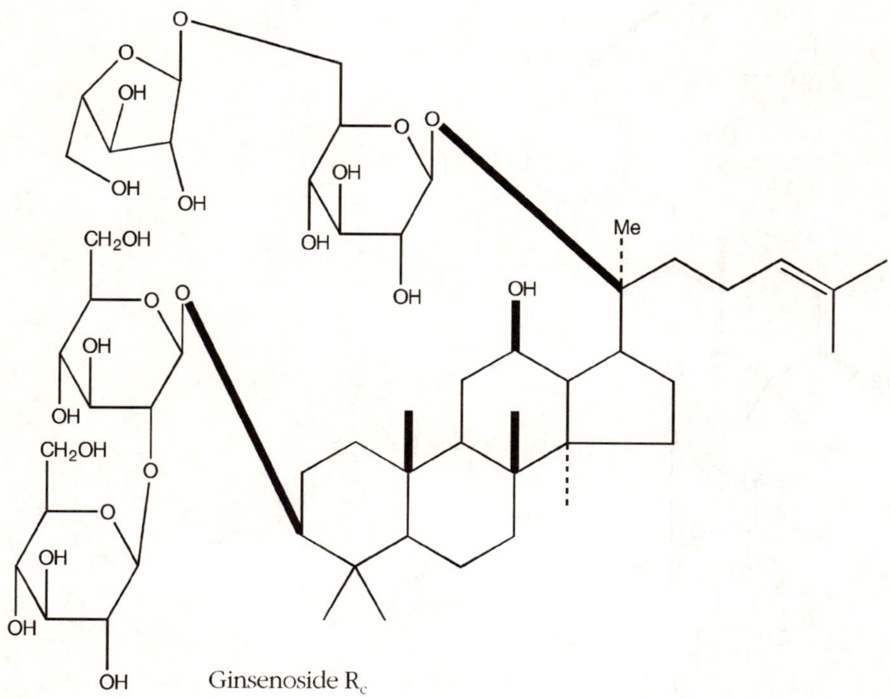


*Figure 5.2* Chemical structure of ginsenoside $R_c$, a protopanaxdiol found in *Radix Ginseng*.

Several examples are listed in the Table 5.1.

*Table 5.1* Protopanaxadiol-type ginsenosides from ginseng

| *Ginsenoside* | *Reference* |
|---|---|
| $Ra_1$ | Besso 1982a, Koizumi 1982 |
| $Ra_2$ | Besso 1982a |
| $Ra_3$ | Matsuura 1984 |
| $Rb_1$ | Kasai 1983, Sanada 1974a |
| $Rb_2$ | Kasai 1983, Sanada 1974a |
| $Rb_3$ | Kasai 1983, Sanada 1978 |
| Rc | Kasai 1983, Sanada 1974a, Luo 1983 |
| Rd | Kasai 1983, Sanada 1974a |
| $Rg_3$ | Kasai 1983, Sanada 1974a |
| Notoginsenoside $R_4$ | Matsuura 1984 |
| $Rh_2$ | Kitagawa 1983a |
| $Rs_1$ | Kasai 1983 |
| $Rs_2$ | Kasai 1983 |
| Quinquenoside $R_1$ | Kasai 1983, Besso 1982b |
| Malonylginsenoside $Rb_1$ | Kitagawa 1983b |
| Malonylginsenoside $Rb_2$ | Kitagawa 1983b |
| Malonylginsenoside Rc | Kitagawa 1983b |
| Malonylginsenoside Rd | Kitagawa 1983b |

(3) Protopanaxtriol type

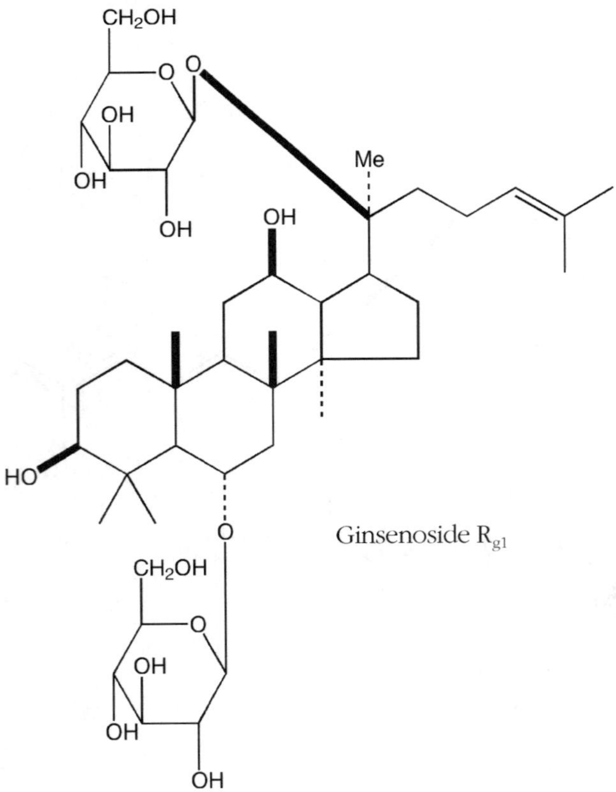


Ginsenoside $R_{g1}$

*Figure 5.3* Chemical structure of ginsenoside $Rg_1$, a protopanaxtriol found in *Radix Ginseng*.

Several examples are listed in the Table 5.2.

*Table 5.2* Protopanaxatriol-type ginsenosides from ginseng

| *Ginsenoside* | *Reference* |
|---|---|
| Re | Kasai 1983, Sanada 1974b |
| Rf | Kasai 1983, Sanada 1974b |
| 20-Glucoginsenoside Rf | Kasai 1983, Sanada 1978 |
| $Rg_1$ | Kasai 1983, Iida 1968, Nagai 1971 |
| $Rg_2$ | Kasai 1983, Sanada 1974b |
| 20 (R) $Rg_2$ | Kitagawa 1983a |
| Notoginsenoside $R_1$ | Kasai 1983 |
| $Rh_1$ | Kitagawa 1983a |

### *B. Volatile components*

*Ginseng* was found to contain a lot of volatile components including a series of sesquiterpenes such as eremphilene, β-gurjunene, *trans*- and *cis*-caryophyllene, ε-muurolene, γ-patchoulene, β-eudesmol, β-farnesene, β-bisabolene, aromadendrene, alloaromadendrene, β-guaiene, γ-elemene, mayurone, pentadecane, 2,5-dimethyltridecane, and palmitic acid (Zhang 1985).

### *C. Polysaccharides*

The polysaccharide preparation of *Ginseng* containing 89 per cent sugars and 5 per cent proteins. The polysaccharide fraction is composed of 80 per cent starch and 20 per cent pectin. The protein fraction contained 15 amino acids such as leucine and arginine (Li 1985).

### *D. Acetylenic compounds*

Panaxynol (Takahashi 1966a; Takahashi 1966b), panaxydol (Poplawski 1980), panaxytriol (Shim 1983), and heptadeca-1-ene-4,6-diyn-3,9-diol were isolated and identified (Dabrowski 1980).

### *E. Peptide glycans*

Isolation of panaxans A, B, C, D, E (Konno 1984), F, G, H (Hikino 1985), I, J, K, L (Oshima 1985), Q, R, S, T, and U (Konno 1985) has been reported.

### *F. Sugars*

*Ginseng* contains a sugar fraction, which comprises two monosaccharides, D-glucose and D-fructose, two disaccharides, sucrose and maltose, and three trisaccharides, maltosyl-β-D-fructofuranose, O-α-D-glucopyranosyl(1→2)-O-β-D-fructofuranosyl(1→2)-β-D-fructofuranose, and O-α-D-glucopyranosyl(1→6)-O-α-D-glucopyranosyl(1→4)-α-D-glucopyranose (Shoji 1985).

### G. *Others*

Besides saponins, *Ginseng* also contains β-sitosterol (Xu 1988), stigmasterol, daucosterol (Lu 1985), campesterol, salicymide (Zhang 1989), adenosine, dencichine, and maltol.

## II. Pharmacology of *Radix Ginseng*

### A. *Effects on CNS*

Ginsenosides, the saponins of *Ginseng*, are active ingredients which exert many biological effects. They have both stimulatory and inhibitory effects on the CNS. It has been reported that intraperitoneal administration of *Ginseng* total saponins not only inhibited hyperactivity and reverse tolerance, but also suppressed postsynaptic dopamine receptor supersensitivity induced by administration of cocaine or nicotine (Kim *et al.* 1995; Kim and Kim 1999).

It was suggested that the learning and memory processes are mediated by the central cholinergic systems. Ginsenosides $Rb_1$ and $Rg_1$ have been shown to prevent scopolamine-induced memory deficits in rats by facilitating acetylcholine release from rat brain hippocampal slices and increasing cholinergic activity (Benishin *et al.* 1992; Yamaguchi *et al.* 1995). Intraperitoneal administration of ginsenoside $Rg_1$ at a dose of 10 mg/kg to rats enhanced the discrimination between two sounds. It also stimulated the exploratory behavior in rats (Shibata *et al.* 1985). Recently, Lee *et al.* (2000) showed that repeated administration of *Ginseng* total saponins ameliorated the impairing effect of ethanol on acquisition, and the effect of the saponins on ethanol-induced amnesia was dependent on the catecholaminergic, but not serotonergic, neuronal activity. Intraperitoneal administration of *Ginseng* saponins with low ratios of protopanaxadiol and protopanaxatriol saponin at doses of 50 and 100 mg/kg respectively, also improved scopolamine-induced learning disability and spatial working memory in mice (Jin *et al.* 1999).

Ginsenosides pretreatment may also prevent ischemic damage to neurons. It has been demonstrated that intracerebroventricular infusion of ginsenoside $Rb_1$, prior to transient forebrain ischemia, precluded significantly ischemia-induced shortening of response latency in a step-down passive avoidance task, and rescued a significant number of hippocampal CA1 neurons from lethal damage (Lim *et al.* 1997; Wen *et al.* 1996).

Ginsenosides may modulate nerve transmission by altering the availability of neurotransmitters. It has been shown that *Ginseng* extract inhibited the uptake of GABA, glutamate, dopamine, noradrenaline, and serotonin in rat brain synaptosomes (Tsang *et al.* 1985). Kimura *et al.* (1994) also demonstrated that ginsenosides $Rb_1$, $Rb_2$, Rc, Re, Rf, and $Rg_1$ competed with agonists for binding to GABA receptors.

### B. *Immunomodulatory effects*

Yun *et al.* (1987) reported that administration of *Ginseng* prevented the depression of natural killer (NK) cell activity induced by injection of carcinogens. The incidence of lung adenoma was lowered following the administration of *Ginseng* in urethane-injected mice. A systematic evaluation of multiple immune system components revealed that *Ginseng* stimulated basal NK cell activity following subchronic exposure and enhanced the recovery of NK function in cyclophosphamide-immunosuppressed mice, but did not further stimulate NK activity in poly I:C treated mice (Kim *et al.* 1990).

The same treatment provided a degree of protection against infection with monocytes, but did not inhibit the growth of transplanted syngeneic tumor cells. Kenarova *et al.* (1990) reported that administration of ginsenoside $Rg_1$ at a dose of 10 mg/kg for three consecutive days before immunization increased the number of spleen plaque-forming cells, the titers of sera hemagglutinins, as well as the number of antigen-reactive T-cells. Ginsenoside $Rg_1$ also increased the number of T-helper cells with respect to the whole T-cell number and the splenocyte natural killer activity. Ginsenoside $Rg_1$ induced an augmentation of the production of IL-1 by macrophages, and exerted a direct mitogenic effect on microcultured thymus cells. Ginsenoside $Rg_1$ also partly restored the impaired immune reactivity by cyclophosphamide treatment. *Ginseng* extract at a concentration of 10 mg/kg significantly enhanced NK-function in peripheral blood mononuclear cells from normal individuals and patients with either chronic fatigue syndrome or acquired immunodeficiency syndrome (Odashima *et al.* 1985).

### *C. Cardiovascular effects*

The cardiovascular effects of *Ginseng* have been studied extensively. Various preparations and concentrations of *Ginseng* have been reported with diverse effects, including the influence on platelet aggregation, and transient vasodilation, in some cases followed by vasoconstriction. It appears that *Ginseng* has both hypotensive and hypertensive effects. The discrepancies of the studies may be due to the difference in manufacturing processes or ginsenoside content of the extracts used. Lee *et al.* (1981) have reported the effect of various extracts of Korean *Ginseng* on cardiovascular function in dogs. Each extract at a dose of 40 mg/kg was administered intravenously to ten dogs under light halothane anesthesia. The administration of ether extract caused a significant decrease in heart rate and central venous pressure. The same treatment regimen with ethanol extract caused a significant decrease in the heart rate and mean arterial pressure. Following the administration of the aqueous extract, the cardiac output, stroke volume, and central venous pressure were significantly decreased, while the total peripheral resistance was significantly increased (Lee *et al.* 1981). The cardiovascular function of *Ginseng* appears to be related to its ability to block $\alpha$-receptors, leading to vasodilation and catecholamine release. However, Gillis (1997) questioned whether some of these complex vascular effects might reflect the qualitative and quantitative heterogeneity of ginsenoside action on different vascular beds that could also be observed *in vitro*. Total *Ginseng* root extract did not alter basal vascular tone of vessels from rabbits, dogs, and humans, but relaxed those vessels that were pre-contracted with either norepinephrine or prostaglandin $F_{2\alpha}$. The author therefore suggested that these actions could reflect the interaction of *Ginseng* with an endogenous vasoactive substance which appeared to be nitric oxide (Gillis 1997). Recently, Maffei Facino *et al.* (1999) showed that oral administration of *Ginseng* extract at a daily dose of 1.6 g/kg for one week prevented the myocardial ischemia/reperfusion damage and the impairment of endothelial functionality induced by reactive oxygen species arising from hyperbaric oxygen exposure. The *in vitro* radical scavenging activity of the *Ginseng* extract seemed to be too weak to explain by itself the cardiac and extra-cardiac protective effects. Therefore, the author suggested that the protective effect of *Ginseng* extract was mediated by the stimulation of endothelial nitric oxide synthase, an indirect antioxidant action of the drug (Maffei Facino *et al.* 1999).

Beside the action on blood pressure, the effect of *Ginseng* on platelet aggregation has been reported. Matsuda *et al.* (1986a) showed that ginsenoside $Rg_2$ strongly inhibited platelet aggregation induced by aggregating agents. Ginsenoside $Rg_2$ also inhibited the conversion of fibrinogen to fibrin induced by thrombin. Another study by Matsuda *et al.* (1986b) showed that ginsenoside $R_0$ prevented disseminated intravascular coagulation induced by infusion of endotoxin or thrombin in rats. Ginsenoside $R_0$ also inhibited the formation of fibrin thrombi in the renal glomeruli in thrombin-induced disseminated intravascular coagulation. The effect of *Ginseng* saponins on aggregation and 5-hydroxytryptamine (5-HT) release with human platelets was reported by Kimura *et al.* (1988). Among the six saponins tested, only ginsenoside $Rg_1$ inhibited adrenaline- and thrombin-induced platelet aggregation and 5-HT release dose-dependently, at concentrations of 5 to 500 mg/ml. Ginsenoside $Rg_1$ had no effect on adrenaline- and thrombin-induced arachidonic acid release and diacylglycerol production. However, it did reduce the elevation of cytosolic free calcium concentration shown in the second phase induced by adrenaline and thrombin, at concentrations ranging from 10 to 500 mg/ml, respectively (Kimura *et al.* 1988). The authors therefore suggested that the inhibitory effects of ginsenoside $Rg_1$ on 5-HT release and aggregation of platelets may be due to the reduction of free calcium elevation at the second phase induced by adrenaline and thrombin. A comparative study on the anti-platelet effect of panaxynol isolated from diethyl ether layer of *Ginseng* extract with those ginsenosides from the butanol layer, showed that panaxynol was the most potent anti-platelet agent in *Ginseng*, and its mechanism of action was mainly due to the inhibition of thromboxane formation (Teng *et al.* 1989).

### D. Hematopoietic effects

The stimulatory effect of *Ginseng* on the hematopoietic function of the bone marrow has been demonstrated by monitoring the levels of blood cells and hemoglobin in normal and anemic animals. An *in vitro* study showed that total saponins of *Ginseng* enhanced the proliferation of progenitor cells from normal individuals (Gao *et al.* 1992). The cell number in bone marrow culture of BFU-E, CFU-E, and CFU-GM was increased by 37.8, 31.4, and 33.3 per cent respectively. The *Ginseng* saponins were also effective in bone marrow culture from some patients with aplastic anemia who are sensitive to methyltestosterone, but not in culture from patients with immune-mediated and stem cell-decreased aplastic anemia.

### E. Effects on endocrine system

Among the diverse effects of *Ginseng*, endocrine effect is the most clear-cut, particularly the stimulatory effect on the hypothalamopituitary suprarenal cortical system (Huang 1998a). Administration of *Ginseng* to rats decreased the vitamin C level of the adrenal cortex, but this effect was not observed in hypophysectomized animals. It was shown that the decrease of vitamin C was mediated by the stimulatory effect of adrenal-corticotropic hormone (ACTH) on the pituitary gland. The action of *Ginseng* on the adrenal cortex was corroborated by the study of Luo *et al.* (1993). Intragastrical or intraperitoneal administration of ginsenoside $Rb_1$ to mice under swim stress at a dose of 10 mg/kg, completely antagonized the immunosuppression induced by swim stress, and further increased the level of serum corticosterone in the mice. Fulder (1981) also showed that the binding of corticosteroid to certain brain regions was increased in adrenalectomized rats treated with *Ginseng* saponin.

*Ginseng* has been reported to reduce blood sugar levels in normal and alloxan-induced hyperglycemic mice by increasing hepatic lipogenesis and glycogen storage (Oshima *et al.* 1985). Sotaniemi *et al.* (1995) suggested that *Ginseng* may be a useful therapeutic adjunct in the management of non-insulin-dependent diabetes mellitus (NIDDM) patients. Treating 36 NIDDM patients with *Ginseng* at a daily dose of 200 mg/person for 8 weeks improved mood and psychophysical performance in diabetic patients. It also led to beneficial changes in daily life, habits, and diet, as indicated by increased physical activity and reduced body weight. Fasting blood glucose was reduced, and glucose response to an oral glucose tolerance test was improved, a reflection of good glycemic control as evidenced by a decrease in glycosylated hemoglobin (Sotaniemi *et al.* 1995).

However, contradictory results have been reported on the effect of *Ginseng* on sexual development. Early studies showed that administration of *Ginseng* extract to castrated mice would neither prevent the atrophy of the male prostate and seminal vesicles nor promote pubertal development in female mice (Huang 1998a). However, other studies with young rodents showed that *Ginseng* extract produced a stimulatory effect on sexual function. Such stimulation was not observed in hypophysectomized rats. It implied that the action of *Ginseng* on the sex organs was mediated by a stimulation of the hypothalamopituitary axis through increasing the secretion of gonadotropins (Huang 1998a).

With an *in vitro* tissue superfusion technique, Chen and Lee (1995) demonstrated that ginsenosides at concentrations ranging from 250 to 750 mg/ml caused a dose-dependent relaxation of corpus cavernosum. Acetylcholine-induced relaxation was increased in the presence of ginsenosides at a concentration of 250 mg/ml. Ginsenosides at 100 mg/ml significantly enhanced the tetrodotoxin-sensitive relaxation of corpus cavernosum elicited by transmural nerve stimulation. The authors suggested that ginsenosides may cause the release of nitric oxide from endothelial cells, and enhance nitric oxide release from endothelial cells elicited by other vasoactive substances and from perivascular nitrergic nerves in the corpus cavernosum (Chen and Lee 1995). A recent study showed that long-term administration of *Ginseng* in rabbits enhanced erectile capacity that was mediated by endothelium-derived relaxing factor and peripheral neurophysiologic enhancement (Choi *et al.* 1999).

### F. *Effect on metabolism*

#### *(1) Carbohydrate metabolism*

*Ginseng* is used in Chinese medicine to reduce blood sugar level in diabetic patients. *Ginseng* polypeptide isolated from the root of *Panax ginseng* has been found to decrease the level of blood sugar and liver glycogen when injected intravenously to rats at doses of 50–200 mg/kg (Wang *et al.* 1990). The polysaccharides increased the content of pyruvic acid, but decreased the content of lactic acid by decreasing the activity of lactate dehydrogenase in the liver. *Ginseng* polysaccharides also accelerated oxidative phosphorylation of carbohydrate by the enhancement of succinate dehydrogenase and cytochrome oxidase activities (Yang *et al.* 1990). Recently, Onomura *et al.* (1999) found that *Ginseng* treatment inhibited absorption of glucose or maltose in rat and human duodenal mucosa, and at the same time increased duodenal muscle movement.

*(2) Lipid metabolism*

It has been reported that treating rats intraperitoneally with ginsenosides $R_c$ at 50 mg/kg increased epididymal synthesis of fat and intestinal fat synthesis by three- and six-fold respectively, in rats pretreated with acetic acid. Many ginsenosides, such as $Rb_1$, $Rb_2$, Rc, Rd, Re, $Rg_1$, $Rg_2$ and $Rh_1$ have been shown to be able to antagonize the lipolytic effects of ACTH or insulin. Several studies have showed that oral administration of ginsenosides would significantly reduce the level of cholesterol in plasma and liver. Yang *et al.* have demonstrated that ginsenoside treatment combined with aerobic exercise could lower serum cholesterol and triglycerides in diet-induced hyperlipidemia mice (Yang *et al.* 1999).

*(3) RNA and DNA metabolism*

It has been reported that *Ginseng* treatment could increase serum proteins by enhancing RNA biosynthesis and incorporating amino acids into the nuclei of hepatic and renal cells. An intraperitoneal injection of ginsenosides $Rb_2$, Rc, Re and Rg at doses from 5 to 10 mg/kg increased DNA synthesis in bone marrow cells. RNA, protein and lipid synthesis were also increased. The direct addition of a mixture of ginsenosides ($Rb_1$, $Rb_2$ and Rc) also enhanced DNA synthesis (Yamamoto *et al.* 1978).

### *G. Antineoplastic effects*

*In vitro* and *in vivo* studies have suggested that *Ginseng* may prevent or ameliorate various cancers. Yun (1996, 1999) reported that prolonged administration of red *Ginseng* extract inhibited the incidence of tumors induced by 9,10-dimethyl-1,2-benzanthracene, urethane, and aflatoxin B1. Recent investigation showed that red *Ginseng* extract at 50–400 mg/kg could inhibit 7,12-dimethyl benez[a] anthracene-induced skin papilloma in mice, decrease the incidence of papilloma, prolong the latent period of tumor occurrence, and reduce tumor number per mouse in a dose-dependent manner (Xiaoguang *et al.* 1998).

Some studies showed that ginsenosides showed direct cytotoxic and growth inhibitory effects against tumor cells (Wakabayashi *et al.* 1998; Atopkina *et al.* 1999; Lee *et al.* 1999; Liu 2000). Liu *et al.* (2000) showed that ginsenoside $Rg_3$ activated the expression of cyclin-kinase inhibitors, p21 and p27, arrested LNCaP cells at G1 phase, and subsequently inhibited cell growth through a caspase-3-mediated apoptosis mechanism. A similar study showed that the cytotoxic effect of ginsenoside $Rh_2$ in B16 melanoma cells was mediated by the induction of G1 arrest and concomitant suppression of cyclin-dependent kinase-2 activity (Moon *et al.* 2000). Several studies have reported the anti-metastatic effect of ginsenosides (Mochizuhk *et al.* 1995; Wakabayashi *et al.* 1998). Wakabayashi *et al.* (1997) showed that the *in vivo* anti-metastatic effect induced by oral administration of *Ginseng* protopanaxadiol saponins was mediated by their metabolic component M1, and that the growth, invasion and migration of tumor cells were inhibited by M1, but not by ginsenosides. Their further investigations demonstrated that the M1 induced apoptotic cell death in B16 melanoma cells by up-regulating the expression of p27Kip1 and down-regulating the expression of c-Myc and cyclin D1 (Wakabayashi *et al.* 1998). Study of Mochizuki *et al.* (1995) showed that two saponin preparations from red *Ginseng*, 20(R)- and 20 (S)-ginsenoside-$Rg_3$, inhibited lung metastasis produced by B16-BL6 melanoma and colon 26-M3.1 carcinoma in syngeneic mice, and suggested that the anti-metastatic effect of the two saponins was related to inhibition of the adhesion and invasion of tumor cells and anti-angiogenesis activity.

### *H. Antioxidant activities*

Oxygen-derived reactive species play an important role in many human diseases such as ischemia/reperfusion injury in the heart. Since *Ginseng* is an important component of Shengmai San that is used for the treatment of myocardial diseases, many studies attributed the myocardial protective effect of *Ginseng* to its antioxidant activities (Mei *et al.* 1994; Liu *et al.* 1998; Maffei Facino *et al.* 1999). Liu *et al.* (1998) have demonstrated that after global ischemia for 60 minutes followed by 30 minutes of reperfusion in a transplanted heart, superoxide dismutase activity in the myocardium treated with ginsenosides was significantly enhanced when compared with the control group. The levels of oxygen free radicals and malondialdehyde, an index of lipid peroxidation, were markedly decreased in the myocardium treated with ginsenosides (Liu *et al.* 1998). Recent studies suggested that *Ginseng* prevented myocardial ischemia/reperfusion injury by an indirect antioxidant action of the drug through the stimulation of endothelial nitric oxide synthase and the subsequent prevention of reactive oxygen species-induced impairment of the endothelial functionality (Mei *et al.* 1994; Maffei Facino *et al.* 1999).

Besides, some studies have reported the *in vivo* and *in vitro* antioxidant activity of *Ginseng* in other tissues (Zhong and Jiang 1997; Voces *et al.* 1999; Keum *et al.* 2000). Voces *et al.* (1999) showed that prolonged treatment with a standardized *Ginseng* extract significantly increased the hepatic glutathione peroxidase activity and the levels of glutathione in the liver, with a dose-dependent reduction of the thiobarbituric acid reactive substances (TBARS, an indirect index of lipid peroxidation). Zhang *et al.* (1997) have demonstrated that *Ginseng* extract directly inhibited the decomposition of unsaturated fatty acids caused by iron and hydrogen peroxide-induced lipid peroxidation. A recent study reported that heat treatment of *Ginseng* at a temperature higher than that applied to the conventional preparation of red *Ginseng* yielded a mixture of saponins with potent antioxidative properties (Keum *et al.* 2000). The methanol extract of this 'neoginseng' attenuated lipid peroxidation in rat brain homogenates induced by ferric ion or ferric ion supplemented with ascorbic acid. The extract protected against strand scission in phiX174 supercoiled DNA induced by UV photolysis of $H_2O_2$, and was also capable of scavenging superoxide generated by xanthine-xanthine oxidase or by 12-O-tetradecanoylphorbol-13-acetate in differentiated human promyelocytic leukemia cells (Keum *et al.* 2000).

### *I. Adaptogenic effects*

The term adaptogen was used to describe the non-specific 'tonic' effect of a substance (Brekhman and Dardymov 1969). *Ginseng* has been used for several thousand years as a tonic, prophylactic agent, and a 'restorative'; it is also used by athletes as an ergogenic and anti-fatigue agent (Bahrke and Morgan 1994). D'Angelo *et al.* (1986) reported that healthy male volunteers, given a standardized preparation of Korean *Ginseng* for 12 weeks, showed better performance in certain psychomotor functions. A double-blind, randomized, crossover study in 50 healthy male sports teachers showed that *Ginseng* preparation increased the subjects' work capacity by improving muscular oxygen utilization (Pieralisi *et al.* 1991). At the same workload, oxygen consumption, plasma lactate levels, ventilation, carbon dioxide production, and heart rate were decreased in subjects receiving *Ginseng* treatment (Pieralisi *et al.* 1991). The adaptogenic actions of *Ginseng* were more obvious when the tested subject was under stressful conditions (Bahrke and Morgan 1994). Recent study showed that *Ginseng* treatment improved psychomotor performance at rest and during graded exercise in young athletes, as indicated by a shortening of reaction time to multiple-choice questions (Ziemba *et al.* 1999).

The adaptogenic effects of *Ginseng* have been attributed to the increase in hypothalamic-pituitary-adrenal sensitivity (Fulder 1981). It has been reported that an intraperitoneal injection of *Ginseng* increased plasma immunoreactive adrenocorticotropic hormone and corticosterone in rats, and the effects were abrogated by hypophysectomy (Hiai *et al.* 1983). Besides, some studies attributed the adaptogenic effects of *Ginseng* to its action on the CNS (Bhattacharya and Mitra 1991; Kim *et al.* 1992).

### *J. Pharmacokinetics*

Following an oral administration of [$^3$H]ginsenoside $Rg_1$, the bioavailability of the ginsenoside in the blood was 49 per cent and peaked at 2.1 hours (Liu and Xiao 1992). After an intravenous injection of [$^3$H]ginsenoside $Rg_1$ in mice, the radioactivity decreased in a triphasic manner in the blood, and the distribution of radioactivity in tissues followed a decreasing order of kidney, adrenal gland, liver, lungs, spleen, pancreas, heart, testes, and brain (Liu and Xiao 1992). After administration of ginsenoside $Rg_1$ in rats, only trace amounts of ginsenosides were excreted in the urine (Odani *et al.* 1983). Using GC-MS to analyze urine samples of 65 athletes who had ingested *Ginseng* for 10 days prior to urine collection, an aglycone of ginsenosides, 20(S) protopanaxatriol, was detected at a concentration of between 2 and 35 ng/ml in 90 per cent of the samples (Cui *et al.* 1996).

### *K. Toxicity*

The $LD_{50}$ of *Ginseng* root in mice was reported to be 10 to 30 g/kg (Gillis 1997). Tam (1992) showed that the $LD_{50}$ values of purified *Ginseng* saponins in male mice were 270 mg/kg (intravenous injection), 342 mg/kg (intraperitoneal injection), 505 mg/kg (intramuscular injection), 950 mg/kg (subcutaneous injection), and 5 g/kg (oral administration). Clinical study on patients using 3 per cent *Ginseng* tincture up to 100 ml showed slight irritation and excitation. When the dose was doubled, urticaria, itching, headache, dizziness, hemorrhage, and insomnia were observed (Huang 1998a).

## FRUCTUS SCHISANDRAE

*Fructus Schisandrae* (*Wuweizi*) is the fruit of *Schisandra chinensis* (Turcz.) Baill. or *Schisandrae sphenanthera* Rehd et Wils. The fruit has a sweet and sour *taste* and is used as a tonic and sedative in traditional Chinese medicine. Starting from the 1950s, pharmacological effects of *Fructus Schisandrae* on the central nervous system and the cardiorespiratory function have been reported. At the beginning of 1970s, the herb was used clinically for the treatment of chronic viral and chemical hepatitis.

## I. Chemical constituents of *Fructus Schisandrae*

A series of compounds have been isolated from *Fructus Schisandrae*, and they can be classified into the following types.

### *A. Volatile components*

*Fructus Schisandrae* was found to contain a lot of volatile components including α- and β-pinene, camphene, myrcene, limonene, α-terpinene, γ-terpiene, *p*-cymene, thymol methylether, bornyl acetate, citronellyl acetate, linalool, terpinen-4-ol, geraniol, borneol, cuparene, and chanigrenol (Ohta 1968a; Ohta 1968b).

### *B. Lignans*

A series of lignans of the dibenzo[*a*,*c*]cyclooctadiene type have been isolated from *Schisandra* species, some of which showed significant biological and biochemical activities in experimental and clinical studies. Lignans are plant phenols whose structure is represented by the union of two cinnanmic acid residues or their biogenetic equivalents. The first compound of the dibenzo[*a*,*c*]cyclooctadiene type (Figure 5.4) isolated from seed oil of *Schisandra chinensis* was schisandrin. Selected examples of reported compounds in *Fructus Schisandrae* are shown in the Table 5.3.

Several lignans derivatives other than dibenzo[*a*,*c*]cyclooctadiene were also isolated such as pregomisin (Figure 5.4) (Ikeya 1978b).

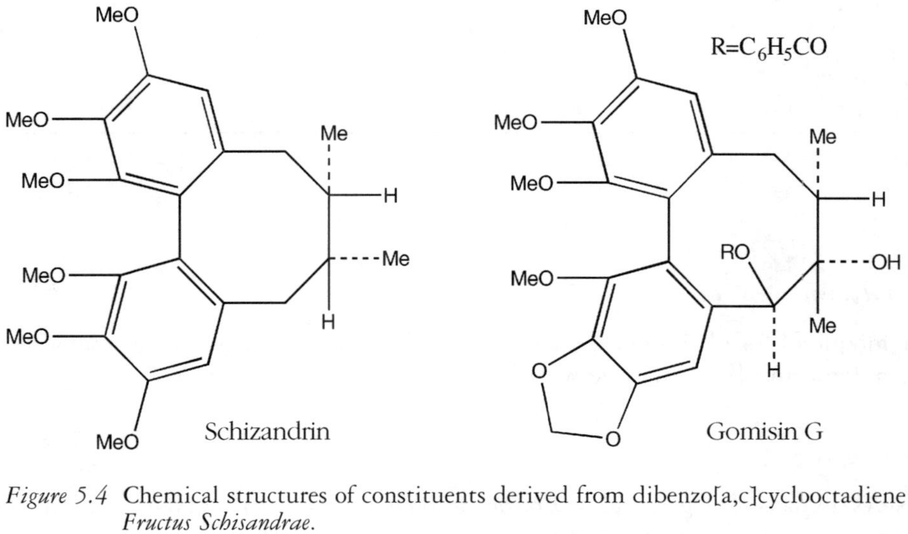


*Figure 5.4* Chemical structures of constituents derived from dibenzo[a,c]cyclooctadiene in *Fructus Schisandrae*.

*Table 5.3* Lignans reported from *Wuweizi*

| *Compound* | *Reference* |
|---|---|
| Schisandrin A | Song 1983, Chen 1976 |
| Schisandrin B | Song 1983, Chen 1976, Ikeya 1982a, Tan 1986 |
| Schisandrin C | Song 1983, Chen 1976, Ikeya 1982b |
| Schisandrol A | Kotchetkov 1961, Ikeya 1979a |
| Schisandrol B | Ikeya 1979a, Taguchi 1975, Taguchi 1977 |
| Schisantherin A | Ikeya 1979a, Taguchi 1975, Taguchi 1977 |
| Schisantherin B | Ikeya 1979a, Taguchi 1975, Taguchi 1977 |
| Isoschisandrin | Ikeya 1988a |
| Gomisin D | Ikeya 1976, Ikeya 1979b |
| Gomisin E | Ikeya 1979c |
| Gomisin F | Ikeya 1979a |
| Gomisin G | Ikeya 1979a |
| Gomisin H | Ikeya 1979d |
| Angeloylgomisin H | Ikeya 1979d, Ikeya 1978a |
| Benzoylgomisin H | Ikeya 1979d, Ikeya 1978a |
| Tigloylgomisin H | Ikeya 1979d, Ikeya 1978a |
| Gomisin J | Ikeya 1978b, Ikeya 1979e |
| Gomisin $K_1$ | Ikeya 1980a |
| Gomisin $K_2$ | Ikeya 1980a |
| Gomisin $K_3$ | Ikeya 1980a, Liu 1985 |
| Gomisin $L_1$ | Ikeya 1982a |
| Gomisin $L_2$ | Ikeya 1982a |
| Gomisin $M_1$ | Ikeya 1982a |
| Gomisin $M_2$ | Ikeya 1982a |
| Gomisin N | Ikeya 1978c, Hikino 1984 |
| Gomisin O | Ikeya 1979c |
| Angeloylgomisin O | Ikeya 1982c |
| Benzoylisogomisin O | Ikeya 1982c |
| Angeloylisogomisin O | Ikeya 1982c |
| Epigomisin O | Ikeya 1979c |
| Angeloylgomisin P | Ikeya 1980b |
| Trgloylgomisin P | Ikeya 1978a, Ikeya 1980b |
| Angeloylgomisin Q | Ikeya 1979f |
| Gomisin R | Ikeya 1982b |
| Gomisin S | Ikeya 1988b |
| Gomisin T | Ikeya 1988b |
| Deoxygomisin A | Hikino 1984 |

### C. Glycosides

Thymoquinol-5-O-β-D-glucopyranoside, thymoquinol-2-O-β-D-glucopyranoside, and kaempferol-3-O-β-ruutinoside were isolated (Yahara 1993).

### D. Others

Besides lignans and glycosides, *Fructus Schisandrae* was also found to contain citral, chlorophyll, protocatechuic acid and vitamin C and E (Yahara 1993).

## II. Pharmacology of Fructus Schisandrae

### A. *Effects on CNS*

Different components isolated from *Schisandra chinensis* have been shown to have a contradictory effect on the CNS. *Fructus Schisandrae* had a stimulatory action in strengthening the spinal reflex and reducing the reflex latency. Volatile oil from *Fructus Schisandrae* could reduce the sleeping time induced by pentobarbital. However, the ethanol extract and the isolated schisandrins significantly prolonged the sleeping time (Niu *et al.* 1983; Song *et al.* 1990). This action has been attributed to the inhibition of microsomal enzymes in the liver.

It has been shown that an intraperitoneal injection of schisandrol A at a dose of 11 mg/kg (1/50 of $LD_{50}$) decreased spontaneous motor activity in mice. The treatment also enhanced the inhibitory effects of chlorpromazine, reserpine and pentobarbital, and antagonized the stimulating effects of amphetamine and caffeine on spontaneous motor activity in mice (Niu *et al.* 1983). Besides, Song *et al.* reported that schisandrin and daucosterol isolated from a methanol extract of *Schisandra chinensis* could elevate the neurotransmitter GABA level in the CNS by inhibiting GABA-degradative enzymes.

### B. *Cardiovascular effects*

The extract of the herb has been reported to induce vasodilation and hypotension in humans and animals. However, the extract increased blood pressure under circulatory failure, and thus it might produce a regulatory effect on blood pressure (Zhu 1998b). A recent study showed that treating rats with schisandrin B at doses of 0.6 and 1.2 mmol/kg for three days protected against myocardial ischemia-reperfusion injury in a dose-dependent manner (Yim and Ko 1999). The myocardial protection was associated with an enhancement in myocardial glutathione antioxidant status, as indicated by significant reductions in glutathione depletion and inhibition of Se-glutathione peroxidase and glutathione reductase activities induced by ischemia-reperfusion in isolated hearts (Yim and Ko 1999).

### C. *Hepatoprotective effect*

The hepatoprotective effect is the most important action of *Schisandra chinensis.* Pretreating experimental animals with the extracts or lignans isolated from the herb protected against hepatic injury induced by a variety of chemical and immunological toxins, as indicated by the significant decreases of plasma alanine aminotransferase activity and necrosis of hepatocytes (Nagai *et al.* 1989; Lu and Liu 1991; Mizoguchi *et al.* 1991a, 1991b; Yamada *et al.* 1993). The hepatoprotection was at least in part attributed to the induction of hepatic cytochrome P-450 dependent enzymes and glutathione *S*-transferases for the detoxification reactions. In addition, the ability to inhibit lipid peroxidation under oxidative stress may also contribute to the hepatoprotective action of the lignans against free-radical mediated damage in the liver (Liu *et al.* 1992; Lu and Liu 1992; Zhang *et al.* 1992).

The lignans have been found to enhance the proliferation of endoplasmic reticulum and induce hypertrophy in the liver. The syntheses of hepatic proteins and microsomal cytochrome P-450 content were also increased by treating animals with the lignans

(Liu *et al.* 1980; Liu *et al.* 1981). Oral administration of gomisin A, a lignan component of *Fructus Schisandrae*, enhanced liver functions, as indicated by the increases in bile secretion and hepatic secretion of bromosulfophthalein, presumably through the increase of liver blood flow (Takeda *et al.* 1988). It has been demonstrated that treating rats with gomisin A accelerated both the proliferation of hepatocytes and the recovery of liver function after partial hepatectomy (Takeda *et al.* 1986). An oral administration of gomisin A to rats 30 minutes before partial hepatectomy significantly increased the mitotic index and the level of DNA synthesis. The treatment stimulated liver regeneration after partial hepatectomy by enhancing ornithine decarboxylase activity, which is an important biochemical event in the early stages of liver regeneration (Kubo *et al.* 1992).

### *D. Antioxidant activities*

Free radical scavenging activity of the lignans isolated from *Fructus Schisandrae* has been extensively investigated in a number of *in vitro* assay systems using microsomes, mitochondria, and homogenate prepared from liver, brain, heart, and kidney (Lin *et al.* 1991; Liu *et al.* 1992; Lu and Liu 1992; Zhang *et al.* 1992). The scavenging activity was demonstrated either directly by using electron magnetic resonance measurement of free radicals (Li *et al.* 1990; Lin *et al.* 1990) or indirectly by measuring the formation of malondialdehyde and the membrane integrity (Zhang *et al.* 1989). In all cases, the lignans were found to be more potent than vitamin E in the inhibition of lipid peroxidation (Lin *et al.* 1990).

However, the antioxidant potential of *Fructus Schisandrae* is found to be mainly due to its modulation on *in vivo* antioxidant mechanisms as implicated in animal studies (Ko *et al.* 1995; Ip *et al.* 1995, 1996, 1997; Ip and Ko 1996). In a study examining the effect of the herb on hepatic glutathione status, it was shown that treating rats with a petroleum ether extract of *Fructus Schisandrae* for three days caused a dose-dependent enhancement on hepatic glutathione status, as indicated by increases in hepatic glutathione level and activities of hepatic glutathione reductase and glucose-6-phosphate dehydrogenase (Ko *et al.* 1995). The treatment also decreased the susceptibility of liver homogenate to glutathione depletion induced by t-butylhydroperoxide. The enhancement in hepatic glutathione status has been ascribed to the decreases of cellular damage and lipid peroxidation in liver of carbon tetrachloride ($CCl_4$)-treated rats, as assessed by plasma activity of alanine aminotransferase and hepatic level of malondialdehyde, respectively (Ko *et al.* 1995a). The results suggested that the hepatoprotective effect of the lignoid extract may involve the facilitation of glutathione regeneration catalyzed by glutathione reductase. A subsequent study showed that pretreating rats with the lignoid extract caused a significant enhancement of hepatic glutathione regeneration capacity in $CCl_4$-treated animals (Ko *et al.*1995b).

Using an animal model of $CCl_4$-induced hepatotoxicity, the activity-directed fractionation of the extract of *Fructus Schisandrae* has led to the isolation of schisandrin B as the major active component (Ip *et al.* 1995). Schisandrin B pretreatment caused a dose-dependent protection against $CCl_4$-induced hepatocellular damage. Schisandrin B treatment caused linear increases in activities of hepatic glutathione-S-transferase and glutathione reductase (Ip *et al.* 1995). A subsequent study found that schisandrin B protected against $CCl_4$ hepatotoxicity by enhancing the mitochondrial glutathione redox status in mouse liver (Ip *et al.* 1996). A comparative study between schisandrin

B and butylated hydroxytoluene, a synthetic phenolic antioxidant, suggested that the ability to sustain the hepatic mitochondrial glutathione level and the hepatic ascorbic acid and α-tocopherol levels may represent a crucial antioxidant action of schisandrin B in protecting against $CCl_4$-induced hepatocellular damage (Ip and Ko 1996). A study using schisandrins A, B and C on $CCl_4$ toxicity showed that the methylenedioxy group of the lignan molecule was an important structural determinant in enhancing hepatic mitochondrial glutathione status particularly under oxidative stress (Ip *et al.* 1997).

### *E. Anticarcinogenic effects*

The lignans isolated from *Fructus Schisandrae* were found to be able to inhibit tumor formation induced by chemical carcinogens. The mutagenic and oncogenic effects of carcinogens require metabolic activation to their ultimate forms by microsomal cytochrome P-450 associated monooxygenases. An *in vitro* study showed that several lignans from the herb inhibited hepatic microsomal hydroxylation of benzo[a]pyrene and demethylation of aminopyrine (Liu and Lesca 1982). Schisandrin B and schizandrol B also decreased mutagenicity of benzo[a]pyrene in an Ames test.

Yasukawa *et al.* (1992) showed that gomisin A, gomisin J, and schisandrin C inhibited the inflammatory activity induced by 12-O-tetradecanoylphorbol-13-acetate in mice. In addition, treating mice with gomisin A at a dose of 5 μmol/mouse markedly suppressed the promotion effect of 12-O-tetradecanoylphorbol-13-acetate on skin tumor formation in mice following initiation with 7,12-dimethylbenz[a]anthracene. It was suggested that the inhibition of tumor promotion by gomisin A was due to its anti-inflammatory activity (Yasukawa *et al.* 1992). Another study by Ohtaki *et al.* (1994) showed that gomisin A inhibited the increases of glutathione S-transferase placental form (a marker enzyme of preneoplasm) induced by 3′-methyl-4-dimethylaminoazobenzene (3′-MeDAB) in rats. It also reduced the population of diploid nuclei (a proliferative state of hepatocytes) in the liver of 3′-MeDAB-treated rats. The authors suggested that gomisin A inhibited the hepatocarcinogenesis induced by 3′-MeDAB through an enhancement of the excretion of the carcinogen from the liver and reversing the normal cytokinesis (Ohtaki *et al.* 1994).

### *F. Effects on physical exercise*

Like *Ginseng*, *Schisandra chinensis* has been used as an adaptogen which can counteract fatigue and improve physical performance. Treating Polo horses with an ethanol extract of *Fructus Schisandrae* reduced the increase of heart rate induced by exercise (Ahumada *et al.* 1989). After the race, the horses treated with the extract showed a quick recovery of respiratory frequency and a significant reduction of the concentration of lactate in plasma. The horses treated with the extract also showed better performance in the race. A similar study showed that treating race and spring horses with an ethanol extract of *Fructus Schisandrae* at a single dose of 12 g 30 minutes prior to exercise reduced heart rate and respiratory frequency of the horses at different time intervals after the race (Hancke *et al.* 1994). The beneficial effect of the extract was suggested to be mediated by facilitating the recuperation of cardiovascular, respiratory and metabolic systems in horses subjected to different kinds of exercise. A treadmill running study showed that pretreating rats with a lignan-enriched extract of *Fructus Schisandrae* protected against physical exercise-induced muscle damage (Ko *et al.* 1996). Interestingly, the muscle protection afforded by *Fructus Schisandrae* pretreatment was associated with a significant enhancement in hepatic antioxidant status, as assessed by the levels of glutathione and malondialdehyde (Ko *et al.* 1996).

### *G. Pharmacokinetics*

After oral administration of schisandrin at a dose of 15 mg to healthy male subjects, the average value of the maximum plasma concentration of schisandrin was 96.1 ng/ml. The plasma concentration of this substance could be monitored for 8 h after administration (Ono *et al.* 1995).

After intravenous administration of gomisin at doses of 1.6, 4.0 and 10 mg/kg of body weight, the serum concentration of the drug decreased biphasically, and the terminal elimination half-life at each dose was about 70 minutes (Matsuzaki *et al.* 1991). Dose-dependency was observed for the area under the concentration-time curve (AUC). The serum concentration gomisin A increased rapidly, and reached a maximum within 15 to 30 min when administered orally. The $C_{max}$ and the AUC values were not exactly dose-dependent, but the values increased with a dose-up of gomisin A. Matsuzaki *et al.* showed that the biotransformation of gomisin A to Met. B was very rapid when administered both intravenously and orally. The AUC value of Met. B after oral administration of gomisin A at a dose of 1.6 mg/kg was relatively larger than at any other dosages. The authors suggested that gomisin A extensively underwent the first pass effect in rats. In addition, it was found that more than 80 per cent of gomisin A was bound with rat serum protein *in vitro* and *in vivo* (Matsuzaki *et al.* 1991).

### *H. Toxicity*

The toxicity of *Fructus Schisandrae* and its lignans were reported to be low. No death was observed in mice given 5 g/kg herb intragastrically. Luan and He (1992) reported that the $LD_{50}$ of the herb in mice was 7.4 g/kg (intraperitoneal injection). However, some components of *Fructus Schisandrae* are very toxic to insects, and can be used as insecticides. Miyazawa *et al.* (1998) isolated two insecticidal lignans, gomisin B and gomisin N, from the n-hexane extract of *Fructus Schisandrae*, and showed that the $LC_{50}$ values of gomisin B and gomisin N were 0.031 and 0.125 mmol/ml, respectively.

## RADIX OPHIOPOGONIS

*Radix Ophiopogonis* (*Maidong*) is the root tuber of *Ophiopogon japonicus* (Thunb.) Ker-Gawl. The root is sweet with a slightly bitter *taste*. In traditional Chinese medicine, it is used for nourishing the *stomach*, promoting the production of *body fluid*, and nourishing the *lung*. It is also used for nourishing the *heart* and the treatment of angina pectoris.

### I. Chemical constituents of *Radix Ophiopogonis*

*Maidong* was found to have a complex chemical profile, which can be classified into the following types.

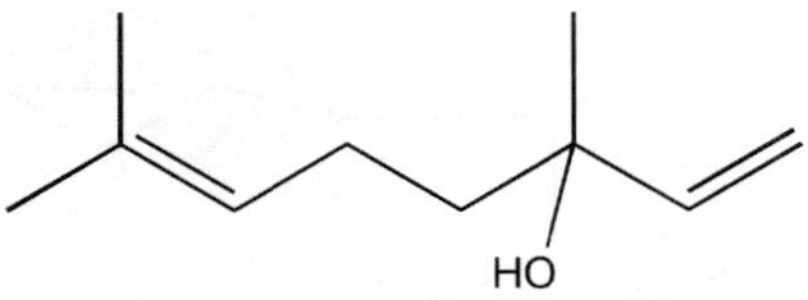


Linalool

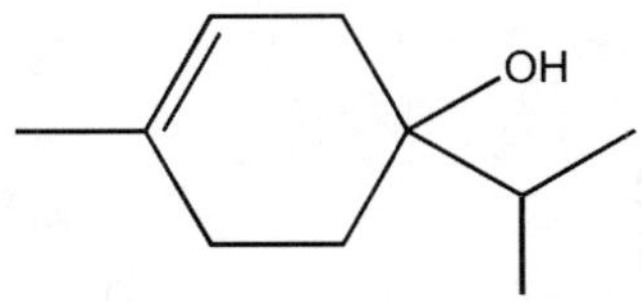


Terpinen-4-ol

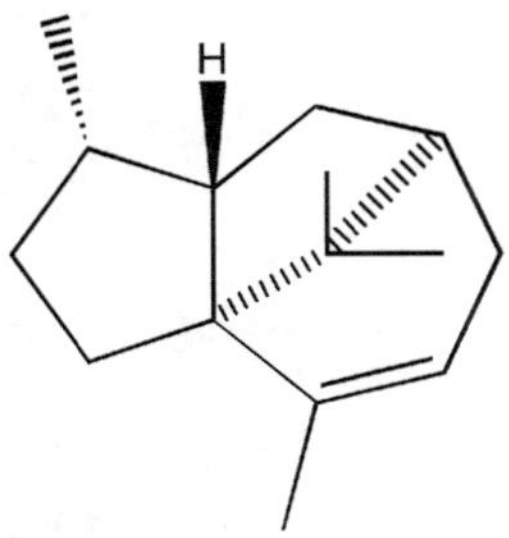


α-Patchoulene

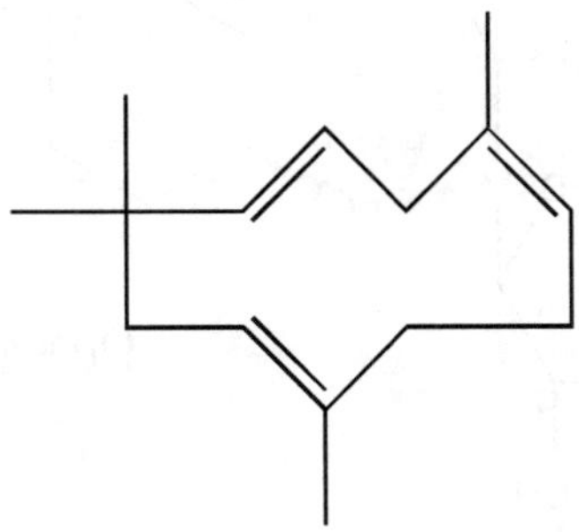

Humulene

*Figure 5.5* Chemical structures of volatile oils found in *Radix Ophiopogonis*.

### A. *Volatile oils (Figure 5.5)*

The volatile oil was found to contain camphor, linalool, terpinen-4-ol, β-patchoulene, longifolene, cyperene, α-humulene, guaiol, α-patchoulene, and jasmololone (Zhu 1991).

### B. *Saponins (Figure 5.6)*

*Maidong* is rich in saponins such as ophiopogonin A, B, C and D (Yang 1987b) where the aglycone is ruscogenin, and ophiopogonin B′, C′ (Tada 1972) and D′ (Tada 1972; Yang 1987a) where the aglycones are diosgenin, diosgenin-3-O-[α-L-rhamnopyranosiyl (1→2)] [(3-O-acetyl)-β-D-xylopyransoyl (1→3)]-β-D-glucopyranoside (Yang 1987a) and diosgenin-3-O-[(2-O-acetyl)-α-L-rhamnopyransoyl (1→2)] [β-D-xylopyranosyl (1→3)]-β-D-glucopyranoside (Tada 1972).

### C. *Flavanoids*

*Maidong* contains homoisoflavanoids such as methylophiopoganone A and B, ophiopogonanone A, 5,7-dihydroxy-6-aldehydo-8-methyl-3-(3,4-methylenedioxylbenzyl)-4-chromanone, 5,6-dihydroxy-6-aldehydo-8-methyl-3-(4-methoxybenzyl)-4-chromanone, 5-hydroxy-6-aldehydo-7-methoxy-8-methyl-3-(3,4-methylenedioxybenzyl)-4-chromanone (Kaneda 1983), 5-hydroxy-6-aldehydo-7-methoxy-8-methyl-3-(4-methoxybenzyl)-4-chromanone (Kaneda 1983; Zhu 1989), 5,7-dihydroxy-6-methyl-3-(4-methoxybenzyl)-4-chromanone (Kaneda 1983), 6-aldehydo-isoophiopogone B, 6-aldehydo-isoophiopogone A (Zhu 1987), and 5,7-dihydroxy-6,8-dimethyl-3-(3,4-methylenedioxybenzyl)-4-chromone (Liu 1991).

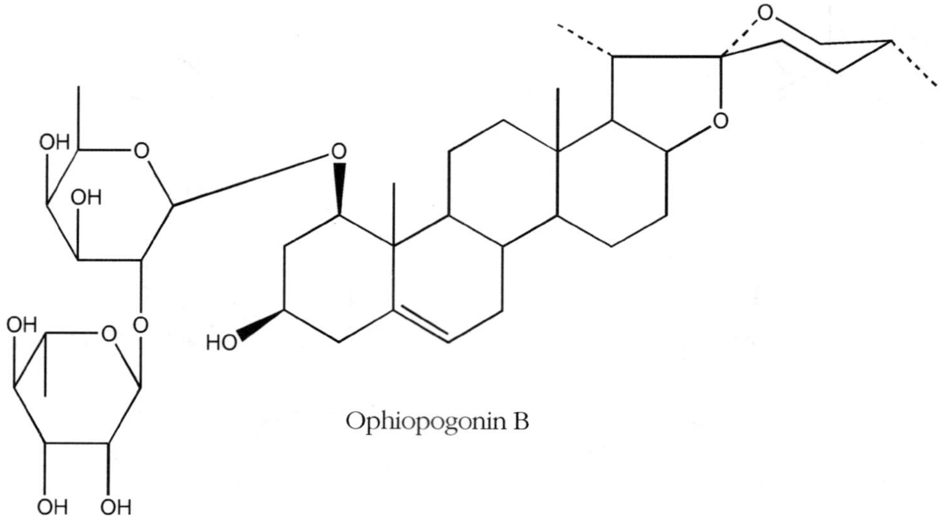


Ophiopogonin B

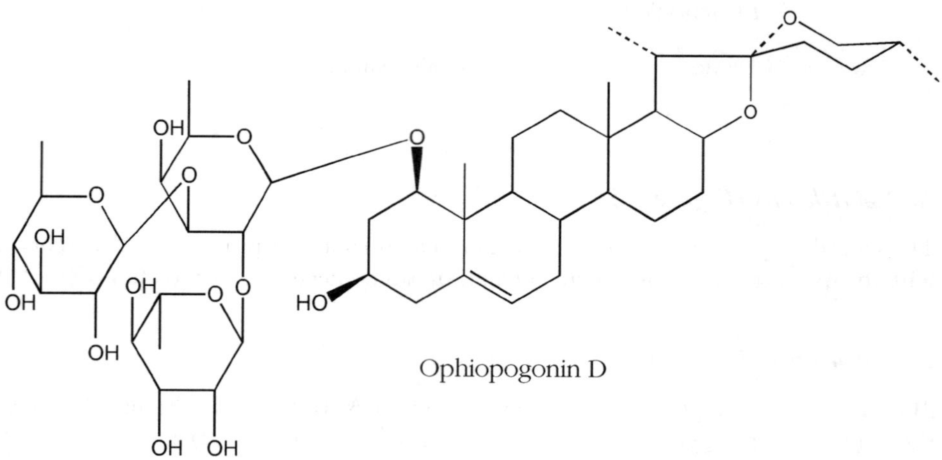


Ophiopogonin D

*Figure 5.6* Chemical structures of saponins found in *Radix Ophiopogonis.*

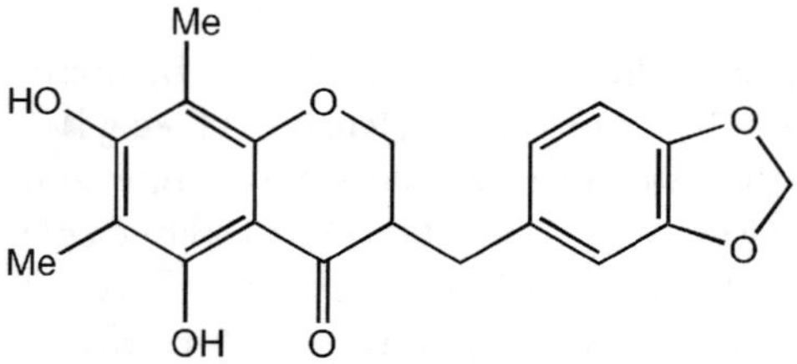


Methylophiopogonanone A

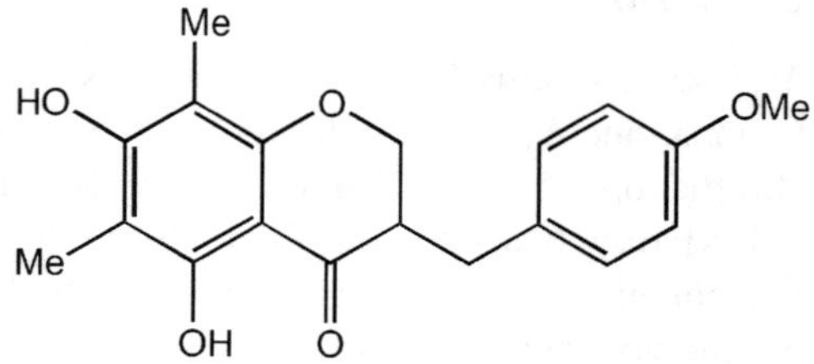


Methylophiopogonanone B

*Figure 5.7* Chemical structures of flavanoids found in *Radix Ophiopogonis.*

### *D. Others*

Besides saponins and flavanoids, *Maidong* also contains β-sitosterol, stigmasterol, β-sitosterol-β-D-glucoside (Zhu 1989), borneol glucoside (Kaneda 1983), borneol-glucoside-(6,1)-apioside (Kaneda 1983; Zhu 1989), and vitamin A.

## II. Pharmacology of Radix Ophiopogonis

### *A. Cardiovascular effects*

It has been reported that *Ophiopogon japonicus* increased coronary blood flow, produced a strophanthin-like inhibitory action on $Na^+/K^+$ ATPase and increased myocardial contractility (Huang 1998b). Clinical trials on 101 cases of coronary disease and angina pectoris found that treating the patients with *Ophiopogon japonicus* showed an improvement in symptoms and ECG pattern (Huang 1998b).

*In vitro* study on cultured human and mouse myocardial cell lines (Dong *et al.* 1995) demonstrated that *Ophiopogon japonicus* significantly enhanced growth metabolism and prolonged the survival time of these cells. The herb also increased the occurrence of spontaneous contraction of mouse myocardial cells (Dong *et al.* 1995).

Chen *et al.* (1990) showed that *Ophiopogon* total saponins extracted from the root of *Ophiopogon japonicus* prevented and antagonized arrhythmias induced by chloroform, epinephrine, $BaCl_2$ and aconitine in dogs. The saponins effectively reduced the incidence of ventricular arrhythmia produced by ligation of the left anterior descending coronary artery, but without producing any changes in the hemodynamic indices. By means of contact electrode and intracellular microelectrode techniques, the saponins were found to shorten action potential duration (APD) at 10, 50 and 90 per cent of repolarization. The saponins also decreased action potential amplitude (APA) and maximum velocity of depolarization (Vmax). In addition, the authors demonstrated that *Ophiopogon* total saponins increased the effective refractory period (ERP)/APD ratio, and prevented or abolished the arrhythmokinesis provoked by ouabain and aconitine. Therefore, Chen *et al.* (1990) suggested that the anti-arrhythmic properties of the saponins were mediated by the blocking of sodium and calcium channels.

### *B. Anti-hypoxic effect*

It has been reported that treating mice with ophiopogonin C at a dose of 20 mg/kg prolonged the survival time of the animals under hypoxic condition. A recent study showed that an injection of alcoholic extract of *Radix Ophiopogon japonicus* at doses of 12.5 and 25 g/kg increased anoxia tolerance in mice by 15.4 and 31.7 per cent, respectively (Li and Zhang 2000).

### *C. Immunomodulatory effects*

Yu *et al.* (1991) showed that treating mice with polysaccharides isolated from *Radix Ophiopogon japonicus* increased the weight of the spleen in animals. After an intravenous injection of charcoal particles, the rate of clearance was enhanced in mice pretreated with the polysaccharides. *Ophiopogon japonicus* also promoted the production of serum specific antibody hemolysin, antagonized leukopenia caused by cyclophosphamide and enhanced the hemagglutination rate in rabbits (Yu *et al.* 1991).

### D. *Hypoglycemic effect*

The hypoglycemic action of polysaccharides isolated from *Radix Ophiopogonis* was studied in normal and alloxan-diabetic mice. In normal mice, these polysaccharides (100 mg/kg, p.o.) significantly lowered blood sugar by 54 per cent at 11 h after administration. In alloxan-diabetic mice, the polysaccharides (200 mg/kg, p.o.) demonstrated marked hypoglycemic actions from 4 h and up to 24 h after the administration (Zhang and Wang 1993).

### E. *Antitussive effects*

Ophiopogonin, a steroid saponin from *Radix Ophiopogonis*, showed strong antitussive effects on bronchitic animals (Miyata *et al.* 1991).

A decrease in neutral endopeptidase activity could increase coughing-inducing substances such as substance P, which stimulates C-fiber endings and induces coughing. Treating animals with ophiopogonin has been shown to prevent the decrease in neutral endopeptidase activity (Miyata *et al.* 1992). Inhalation of substance P, capsaicin, or neurokinin A could induce coughing in normal and bronchitic guinea pigs. The coughing was strongly suppressed by oral administration of 0.5 or 1.0 mg ophiopogonin (Miyata *et al.* 1992). Treating guinea pig with ophiopogonin also suppressed coughing induced by endogenous tachykinin after the administration of phosphoramidon, an inhibitor of enkephalinase and neutral endopeptidase (Miyata *et al.* 1992).

### F. *Toxicity*

Intravenous administration of 1 ml of *Radix Ophiopogon japonicus* extract to mice (corresponding to 2 g of the herbs and 100–12500 times the dosage of man) produced no mortality or toxic reactions. The $LD_{50}$ value of intraperitoneal injection of *Ophiopogon japonicus* in mice was 20.6 g/kg (Zhu 1998a).

Using the micronucleus test of mouse bone marrow cell, Liu and Wu (1999) showed that Radix *Ophiopogonis* at doses of 1.7 and 3.4 g/kg decreased micronuclei frequencies induced by cyclophosphamide. Treating the animals with *Radix Ophiopogonis* at a dose of 6.8 g/kg significantly inhibited the micronuclei frequencies. Thus, the author suggested that *Radix Ophiopogonis* had no genetic toxicity, and may be used as an antimutagen at a high dose (Liu and Wu 1999).

## REFERENCES


Ahumada, H., Hermosilla, J., Hola, R., Pena, R., Wittwer, F., and Wegmann, E. (1989) Studies on the effect of Schizandra chinensis extract on horses submitted to exercise and maximum effort. *Phytotherapy Research*, **3**, 175–179.

Atopkina, L.N., Malinovskaya, G.V., Elyakov, G.B., Uvarova, N.I., Woerdenbag, H.J., Koulman, A., Pras, N., and Potier, P. (1999) Cytotoxicity of natural ginseng glycosides and semisynthetic analogues. *Planta Medica*, **65**, 30–34.

Bahrke, M.S. and Morgan, W.P. (1994) Evaluation of the ergogenic properties of ginseng. *Sports Medicine*, **18**, 229–248.

Benishin, C.G. (1992) Actions of ginsenoside Rb1 on choline uptake in central cholinergic nerve endings. *Neurochemistry International*, **21**, 1–5.

Besso, H., Kaisai, R., Saruwatari, Y., Fuwa, T., and Tanaka, O. (1982a) Ginsenoside-$Ra_1$ and ginsenoside-$Ra_2$, new dammarane-saponins of Ginseng root. *Chemical and Pharmaceutical Bulletin*, **30**, 2380–2385.

Besso, H., Kasai, R., Wei, J., Wang, J.F., Saruwatari, Y., Fuwa T., and Tanaka O. (1982b) Further studies on dammarane-saponins of American Ginseng, root of *Panax quinquefolium* L. *Chemical and Pharmaceutical Bulletin*, **30**, 4534–4538.

Bhattacharya, S.K. and Mitra, S.K. (1991) Anxiolytic activity of Panax ginseng roots: an experimental study. *Journal of Ethnopharmacology*, **34**, 87–92.

Brekhman, I.I. and Dardymov, I.V. (1969) Pharmacological investigation of glycosides from Ginseng and Eleutherococcus. *Lloydia*, **32**, 46–51.

Chen, M., Yang, Z.G., Zhu, J.T., Xiao, Z.Y., and Xiao, R. (1990) Anti-arrhythmic effects and electrophysiological properties of Ophiopogon total saponins. *Acta Pharmacologica Sinica*, **11**, 161–165.

Chen, X. and Lee, T.J. (1995) Ginsenosides-induced nitric oxide-mediated relaxation of the rabbit corpus cavernosum. *British Journal of Pharmacology*, **115**, 15–18.

Chen, Y.Y., Shu, Z.B., and Li, L.N. (1976) Studies on Fructus schisandrae. IV. Isolation and determination of the active compounds (in lowering high SGPT levels) of *Schisandra chinensis* Baill. *Scientia Sinica*, **19**, 276–290.

Choi, Y.D., Rha, K.H., and Choi, H.K. (1999) In vitro and in vivo experimental effect of Korean red ginseng on erection. *Journal of Urology*, **162**, 1508–1511.

Cui, J.F., Garle, M., Bjorkhem, I., and Eneroth, P. (1996) Determination of aglycones of ginsenosides in ginseng preparations sold in Sweden and in urine samples from Swedish athletes consuming ginseng. *Scandinavian Journal of Clinical and Laboratory Investigation*, **56**, 151–160.

Dabrowski, Z., Wrobel, J.T., and Wojtasiewicz, K. (1980) Structure of an acetylenic compound from *Panax ginseng*. *Phytochemistry*, **19**, 2464–2465.

D'Angelo, L., Grimaldi, R., Caravaggi, M., Marcoli, M., Perucca, E., Lecchini, S., Frigo, G.M., and Crema, A. (1986) A double-blind, placebo-controlled clinical study on the effect of a standardized ginseng extract on psychomotor performance in healthy volunteers. *Journal of Ethnopharmacology*, **16**, 15–22.

Dong, J.D., Xia, H.Y.T., Yu, X.P., Zhao, W.M., and Zhou, Y.P. (1995) The effects of herbs and chromium on virus infection and growth metabolism in myocardial cells. *Virologica Sinica*, **10**, 104–110.

Fulder, S.J. (1981) Ginseng and the hypothalamic-pituitary control of stress. *American Journal of Chinese Medicine*, **9**, 112–118.

Gao, R.L., Xu, C.L., and Jin, J.M. (261) Effect of total saponins of Panax ginseng on hematopoietic progenitor cells in normal human and aplastic anemia patients. *Chinese Journal of Integrated Traditional and Western Medicine*, **12**, 285–287.

Gillis, C.N. (1997) Panax ginseng pharmacology: a nitric oxide link?. *Biochemical Pharmacology*, **54**, 1–8.

Hancke, J., Rurgos, R., and Wikman, G. (1994) Schizandra chinensis, a potential phytodrug for recovery of sport horses. *Fitoterapia*, **65**, 113–118.

Hiai, S., Yokoyama, H., Oura, H., and Kawashima, Y. (1983) Evaluation of corticosterone secretion-inducing activities of ginsenosides and their prosapogenins and sapogenins. *Chemical and Pharmaceutical Bulletin*, **31**, 168–174.

Hikino H., Kiso Y., Taguchi H., and Ikeya Y. (1984) Validity of the oriental medicines. LX. Liver protective drugs. 11. Antihepatotoxic actions of lignoids from *Schisandra chinensis* fruits. *Planta Medica*, **50**, 213–218.

Hikino, H., Oshima, Y., Suzuki, Y., and Konno, C. (1985) Isolation and hypoglycemic activity of panaxans F, G and H, glycans of *Panax ginseng* roots. *Shoyakugaku Zasshi*, **39**, 331–333.

Huang, K.C. (1998a) Ginseng. In K.C. Huang (ed.), *The pharmacology of Chinese herbs*, CRC Press, Florida, p. 17–44.

Huang, K.C. (1998b) Mar Dong. In K.C. Huang (ed.), *The pharmacology of Chinese herbs, CRC Press*, Florida, p. 107–108.

Iida, Y., Tanaka, O., and Shibata, S. (1968) Studies on saponins of Ginseng: the structure of ginsenoside-Rg1. *Tetrahedron Letters*, 5449–5453.

Ikeya, Y., Taguchi H., and Itaka Y. (1976) The constituents of *Schisandra chinensis* Baill. The structure of a new lignan, gomisin D. *Tetrahedron Letters*, 1359–1362.

Ikeya, Y., Taguchi, H., and Yosioka, I. (1978a) The constituents of *Schisandra chinensis* Baill. The structures of three new lignans, angeloylgomisin H, tigloylgomisin H and benzoylgomisin H and the absolute structure of schisandrin. *Chemical and Pharmaceutical Bulletin*, **26**, 328–332.

Ikeya, Y., Taguchi, H., and Yosioka, I. (1978b) The constituents of *Schisandra chinensis* Baill. The structures of two new lignans, pre-gomisin and gomisin J. *Chemical and Pharmaceutical Bulletin*, **26**, 682–684.

Ikeya, Y., Taguchi, H., Yosioka, I., and Kobayashi, H. (1978c) The constituents of *Schisandra chinensis* Baill. The structure of two new lignans, gomisin N and tigloylgomisin P. *Chemical and Pharmaceutical Bulletin*, **26**, 3257–3260.

Ikeya, Y., Taguchi H., Yosioka, I., and Kobayashi, H. (1979a) The constituents of *Schisandra chinensis* Baill. I. Isolation and structure determination of five new lignans, gomisin A, B, C, F and G, and the absolute structure of schisandrin. *Chemical and Pharmaceutical Bulletin*, **27**, 1383–1394.

Ikeya, Y., Taguchi, H., Yosioka, I., Iitaka, Y., and Kobayashi, H. (1979b) The constituents of *Schisandra chinensis* Baill. II. The structure of a new lignan, gomisin D. *Chemical and Pharmaceutical Bulletin*, **27**, 1395–1401.

Ikeya, Y., Taguchi, H., Yosioka, I., and Kobaysashi, H. (1979c) The constituents of *Schisandra chinensis* Baill. V. The structures of four new lignans, gomisin N, gomisin O, epigomisin O and gomisin E and transformation of gomisin N to deangeloylgomisin B. *Chemical and Pharmaceutical Bulletin*, **27**, 2695–2709.

Ikeya, Y., Taguchi, H., Yosioka, I., and Kobayashi, H. (1979d) The constituents of *Schisandra chinensis* Baill. III. The structures of four new lignans, gomisin H and its derivatives, angeloyl-, tigloyl- and benzoyl-gomisin H. *Chemical and Pharmaceutical Bulletin*, **27**, 1576–1582.

Ikeya, Y., Taguchi, H., Yosioka, I., and Kobayashi, H. (1979e) The constituents of *Schisandra chinensis* Baill. IV. The structures of two new lignans, pre-gomisin and gomisin J. *Chemical and Pharmaceutical Bulletin*, **27**, 1583–1588.

Ikeya, Y., Taguchi, H., and Yosioka, I. (1979f) The constituents of *Schisandra chinensis* Baill. The cleavage of the methylenedioxy moiety with lead tetraacetate in benzene and the structure of angeloylgomisin Q. *Chemical and Pharmaceutical Bulletin*, **27**, 2536–2538.

Ikeya, Y., Taguchi, H., and Yosioka, I. (1980a) The constituents of *Schisandra chinensis* Baill. VII. The structures of three new lignans, (–)-gomisin $K_1$ and (+)-gomisin $K_2$ and $K_3$. *Chemical and Pharmaceutical Bulletin*, **28**, 2422–2427.

Ikeya, Y., Taguchi, H., Yosioka, I., and Kobayashi, H. (1980b) The constituents of *Schisandra chinensis* Baill. VIII. The structures of two new lignans, tigloylgomisin P and angeloylgomisin P. *Chemical and Pharmaceutical Bulletin*, **28**, 3357–3361.

Ikeya, Y., Taguchi, H., and Yosioka, I. (1982a) The constituents of *Schisandra chinensis* Baill. X. The structure of γ-schizandrin and four new lignans, (-)-gomisins $L_1$ and $L_2$, (±)-gomisin M1 and (+)-gomisin $M_2$. *Chemical and Pharmaceutical Bulletin*, **30**, 132–139.

Ikeya, Y., Taguchi, H., and Yosioka, I. (1982b) The constituents of *Schisandra chinensis* Baill. XII. Isolation and structure of a new lignan, gomisin R, the absolute structure of wuweizisu C and isolation of schisantherin D. *Chemical and Pharmaceutical Bulletin*, **30**, 3207–3211.

Ikeya, Y., Ookawa, N., Taguchi, H., and Yosioka, I. (1982c) The constituents of *Schisandra chinensis* Baill. XI. The structures of three new lignans, angeloylgomisin O and angeloyl- and benzoyl-isogomisin O. *Chemical and Pharmaceutical Bulletin*, **30**, 3202–3206.

Ikeya, Y., Taguchi, H., Mitsuhashi, H., and Takeda, S. (1988a) The constituents of *Schizandra chinensis* Baill. XIV. A lignan from *Schizandra chinensis*. *Phytochemistry*, 27, 569–573.

Ikeya, Y., Kanatani, H., Hakozaki, M., Taguchi, H., and Mitsuhashi, H. (1988b) The constituents of *Schizandra chinensis* Baill. XV. Isolation and structure determination of two new lignans gomisin S and gomisin T. *Chemical and Pharmaceutical Bulletin*, **36**, 3974–3979.

Inoue, M., Wu, C.Z., Dou, D.Q., Chen, Y.J., and Ogihara, Y. (1999) Lipoprotein lipase activation by red ginseng saponins in hyperlipidemia model animals. *Phytomedicine*, 6, 257–265.

Ip, S.P., Poon, M.K., Wu, S.S., Che, C.T., Ng, K.H., Kong, Y.C., and Ko, K.M. (1995) Effect of schisandrin B on hepatic glutathione antioxidant system in mice: protection against carbon tetrachloride toxicity. *Planta Medica*, 61, 398–401.

Ip, S.P. and Ko, K.M. (1996) The crucial antioxidant action of schisandrin B in protecting against carbon tetrachloride hepatotoxicity in mice: a comparative study with butylated hydroxytoluene. *Biochemical Pharmacology*, **52**, 1687–1693.

Ip, S.P., Poon, M.K., Che, C.T., Ng, K.H., Kong, Y.C., and Ko, K.M. (1996) Schisandrin B protects against carbon tetrachloride toxicity by enhancing the mitochondrial glutathione redox status in mouse liver. *Free Radical Biology and Medicine*, **21**, 709–712.

Ip, S.P., Ma, C.Y., Che, C.T., and Ko, K.M. (1997) Methylenedioxy group as determinant of schisandrin in enhancing hepatic mitochondrial glutathione in carbon tetrachloride-intoxicated mice. *Biochemical Pharmacology*, **54**, 317–319.

Jin, S.H., Park, J.K., Nam, K.Y., Park, S.N., and Jung, N.P. (1999) Korean red ginseng saponins with low ratios of protopanaxadiol and protopanaxatriol saponin improve scopolamine-induced learning disability and spatial working memory in mice. *Journal of Ethnopharmacology*, 66, 123–129.

Kaneda, N., Nakanishi, H., Kuraishi, T., and Katori, T. (1983) Studies on the Components of Ophiopogon Roots (China). I. *Journal of the Pharmaceutical Society of Japan*, **103**, 1133–1139.

Kasai, R., Besso, M., Tanaka, O., Saruwatari, Y., and Fuwa, T. (1983) Saponins of red Ginseng. *Chemical and Pharmaceutical Bulletin*, **31**, 2120–2125.

Kenarova, B., Neychev, H., Hadjiivanova, C., and Petkov, V.D. (1990) Immunomodulating activity of ginsenoside Rg1 from Panax ginseng. *Japanese Journal of Pharmacology*, **54**, 447–454.

Keum, Y.S., Park, K.K., Lee, J.M., Chun, K.S., Park, J.H., Lee, S.K., Kwon, H., and Surh, Y.J. (2000) Antioxidant and anti-tumor promoting activities of the methanol extract of heat-processed ginseng. *Cancer Letters*, **150**, 41–48.

Kim, H.S., Oh, K.W., Rheu, H.M., and Kim, S.H. (1992) Antagonism of U-50,488H-induced antinociception by ginseng total saponins is dependent on serotonergic mechanisms. *Pharmacology, Biochemistry and Behavior*, 42, 587–593.

Kim, H.S., Kang, J.G., Seong, Y.H., Nam, K.Y., and Oh, K.W. (1995) Blockade by ginseng total saponin of the development of cocaine induced reverse tolerance and dopamine receptor supersensitivity in mice. *Pharmacology, Biochemistry and Behavior*, **50**, 23–27.

Kim, H.S. and Kim, K.S. (1999) Inhibitory effects of ginseng total saponin on nicotine-induced hyperactivity, reverse tolerance and dopamine receptor supersensitivity. *Behavioural Brain Research*, **103**, 55–61.

Kim, J.Y., Germolec, D.R., and Luster, M.I. (1990) Panax ginseng as a potential immunomodulator: studies in mice. *Immunopharmacology and Immunotoxicology*, 12, 257–276.

Kimura, T., Saunders, P.A., Kim, H.S., Rheu, H.M., Oh, K.W., and Ho, I.K. (1994) Interactions of ginsenosides with ligand-bindings of GABA(A) and GABA(B) receptors. *General Pharmacology*, **25**, 193–199.

Kimura, Y., Okuda, H., and Arichi, S. (1988) Effects of various ginseng saponins on 5-hydroxytryptamine release and aggregation in human platelets. *Journal of Pharmacy and Pharmacology*, **40**, 838–843.

Kitagawa, I., Yoshikawa, M., Yashihara, M., Hayashi, T., and Taniyama, T. (1983a) Chemical studies on crude drug procession. I. On the constituents of Ginseng Radix Rubra. *Yakugaku Zasshi*, **103**, 612–622.

Kitagawa, I., Taniyama, T., Hayashi, T., and Yoshikawa, M. (1983b) Malonyl-ginsenosided-$Rb_1$, -$Rb_2$, -Rc and Rd, four new malonylated dammarane-type triterpene oligosaccharides from Ginseng radix. *Chemical and Pharmaceutical Bulletin*, **31**, 3353–3356.

Ko, K.M., Ip, S.P., Poon, M.K., Wu, S.S., Che, C.T., Ng, K.H., and Kong, Y.C. (1995a) Effect of a lignan-enriched fructus schisandrae extract on hepatic glutathione status in rats: protection against carbon tetrachloride toxicity. *Planta Medica*, **61**, 134–137.

Ko, K.M., Mak, D.H.F., Li, P.C., Poon, M.K., and Ip, S.P. (1995b) Enhancement of hepatic glutathione regeneration capacity by a lignan-enriched extract of fructus schisandrae in rats. *Japanese Journal of Pharmacology*, 69, 439–442.

Ko, K.M., Mak, D.H.F., Li, P.C., Poon, M.K.T., and Ip, S.P. (1996) Protective effect of a lignan-enriched extract of Fructus schisandrae on physical exercise induced muscle damage in rats. *Phytotherapy Research*, **10**, 450–452.

Kochetkov, N.K., Khorlin, A.Y., Chizhov, O.S., and Sheichenko, V.I. (1961) Schizandrin, a lignan of unusual structure. *Tetrahedron Letters*, 730–734.

Koizumi, H., Sanada, S., Ida, Y., and Shoji, J. (1982) Studies on the saponins of Ginseng. IV. On the structure and enzymatic hydrolysis of ginsenoside $Ra_1$. *Chemical and Pharmaceutical Bulletin*, **30**, 2393–2398.

Konno C., Sugiyama K., Kano M., Takahashi M., and Hikino H. (1984) Validity of the oriental medicines. LXX. Antidiabetes drugs. 1. Isolation and hypoglycemic activity of panaxans A, B, C, D and E, glycans of *Panax ginseng* roots. *Planta Medica*, **50**, 434–436.

Konno, C., Murakami, M., Oshima, Y., and Hikino, H. (1985) Validity of the Oriental medicines. CVI. Antidiabetes drugs. 19. Isolation and hypoglycemic activity of panaxans Q, R, S, T and U, glycans of *Panax ginseng* roots. *Journal of Ethnopharmacology*, 14, 69–74.

Kubo, S., Ohkura, Y., Mizoguchi, Y., Matsui-Yuasa, I., Otani, S., Morisawa, S., Kinoshita, H., Takeda, S., Aburada, M., and Hosoya, E. (1992) Effect of Gomisin A (TJN-101) on liver regeneration. *Planta Medica*, 58, 489–492.

Lee, D.C., Lee, M.O., Kim, C.Y., and Clifford, D.H. (1981) Effect of ether, ethanol and aqueous extracts of ginseng on cardiovascular function in dogs. *Canadian Journal of Comparative Medicine*, 45, 182–187.

Lee, S.C., Moon, Y.S., and You, K.H. (2000) Effects of red ginseng saponins and nootropic drugs on impaired acquisition of ethanol-treated rats in passive avoidance performance. *Journal of Ethnopharmacology*, 69, 1–8.

Lee, S.J., Sung, J.H., Lee, S.J., Moon, C.K., and Lee, B.H. (1999) Antitumor activity of a novel ginseng saponin metabolite in human pulmonary adenocarcinoma cells resistant to cisplatin. *Cancer Letters*, 144, 39–43.

Li, L.Q. and Zhang, L.J. (2000) Pharmacological action of Ophiopogon japonicus. *Journal of Hebei Traditional Chinese Medicine and Pharmacology*, 15, 34–35.

Li, R.Q. and Zhang, Y.S. (1985) Analysis of the chemical constituents of ginseng polysaccharide. *Chi Trad Herb Drugs*, 16, 389–391.

Li, X.G. and Teng, F.T. (1979) Studies on the triterpenoid in *Panax ginseng*. *Acta Botanica Sinica*, 21, 181–185.

Li, X.J., Zhao, B.L., Liu, G.T., and Xin, W.J. (1990) Scavenging effects on active oxygen radicals by schizandrins with different structures and configurations. *Free Radical Biology and Medicine*, 9, 99–104.

Lim, J.H., Wen, T.C., Matsuda, S., Tanaka, J., Maeda, N., Peng, H., Aburaya, J., Ishihara, K., and Sakanaka, M. (1997) Protection of ischemic hippocampal neurons by ginsenoside Rb1, a main ingredient of ginseng root. *Neuroscience Research*, 28, 191–200.

Lin, T., Liu, G., and Pan, Y. (1992) Protective effect of schisanhenol against oxygen radical induced mitochondrial toxicity on rat heart and liver. *Biomedical and Environmental Sciences*, 5, 57–64.

Lin, T.J., Liu, G.T., Li, X.J., Zhao, B.L., and Xin, W.J. (1990) Detection of free radical scavenging activity of schisanhenol by electron spin resonance. *Acta Pharmacologica Sinica*, 11, 534–539.

Lin, T.J., Liu, G.T., Pan, Y., Liu, Y., and Xu, G.Z. (1991) Protection by schisanhenol against adriamycin toxicity in rat heart mitochondria. *Biochemical Pharmacology*, 42, 1805–1810.

Liu, B. and Wu, G.H. (1999) Genetic toxicity of Radix Ophiopogonis. *Journal of Norman Bethune University of Medica Sciences*, 25, 129–130.

Liu, C.J., Zeng, Q., Liu, D., Zhang, J., and Yang, Y.D. (1991) Studies on the Constituents of Homoisoflavonoids of *Ophiopogon japonicus* of Hang. *Chinese Traditional and Herbal Drugs*, 22, 60.

Liu, C.X. and Xiao, P.G. (1992) Recent advances on ginseng research in China. *Journal of Ethnopharmacology*, 36, 27–38.

Liu, G.T., Bao, T.T., Wei, H.L., and Song, Z.Y. (1980) Induction of hepatocyte microsomal cytochrome P-450 by Schizandrin B in mice. *Acta Pharmaceutica Sinica*, 15, 206–211.

Liu, G.T. and Wei, H.L. (1985) Induction of hepatic microsomal monooxygenases by schisandrol in rats. *Acta pharmacologica Sinica*, 6, 41–44.

Liu, K., Abe, T., Sekine, S., Goto, Y., Iijima, K., Kondon, K., Matsukawa, M., Tian, J., Wu, W., Zhang, B., Chen, L., Zhang, H., Zhang, X., Zhao, H., and Song, X. (1998) Experimental study on the scavenging effects of ginsenosides on oxygen free radicals using model of heterotopic heart transplantation in rats. *Annals of Thoracic and Cardiovascular Surgery*, 4, 188–191.

Liu, K.T., Cresteil, T., Le Provost, E., and Lesca, P. (1981) Specific evidence that schizandrins induce a phenobarbital-like cytochrome P-450 form separated from rat liver. *Biochemical and Biophysical Research Communications*, 103, 1131–1137.

Liu, K.T. and Lesca, P. (1982) Pharmacological properties of dibenzo[a,c]cyclooctene derivatives isolated from Fructus Schizandrae chinensis. I. Interaction with rat liver cytochrome P-450 and inhibition of xenobiotic metabolism and mutagenicity. *Chemico-Biological Interactions*, 39, 301–314.

Liu, W.K., Xu, S.X. and Che, C.T. (2000) Anti-proliferative effect of ginseng saponins on human prostate cancer cell line. *Life Sciences*, 67, 1297–1306.

Lu, H. and Liu, G.T. (1991) Effect of dibenzo[a,c]cyclooctene lignans isolated from Fructus schizandrae on lipid peroxidation and anti-oxidative enzyme activity. *Chemico-Biological Interactions*, 78, 77–84.

Lu, H. and Liu, G.T. (1992) Anti-oxidant activity of dibenzocyclooctene lignans isolated from Schisandraceae. *Planta Medica*, 58, 311–313.

Lu, Y.J., Zeng, X.Y., Ning, L.D., Wang, H.A., and Wang, H.Y. (1985) Studies on the Chemical Constituents of Ginseng. V. Isolation and Identification of Steroid-saponins and Triterpene-saponins of Ginseng. *Journal of Shenyang College of Pharmacy*, 2, 177–184.

Luan, L. and He, M. (1992) Estimation of apparent pharmacokinetic parameters of herbal Houttuyniae and dructus Schizandrae with the method of acute mortality of animals. *Journal of China Medical University*, 21, 183–186.

Luo, S.D., Liu, A.Y., and Liu, Y.Y. (1983) Comparison of active constituents and pharmacological effects of ginseng root and ginseng rhizome. *Chemical and Pharmaceutical Bulletin*, 18, 468–470.

Luo, Y.M., Cheng, X.J., and Yuan, W.X. (1993) Effects of ginseng root saponins and ginsenoside Rb1 on immunity in cold water swim stress mice and rats. *Acta Pharmacologica Sinica*, 14, 401–404.

Maffei Facino, R., Carini, M., Aldini, G., Berti, F., and Rossoni, G. (1999) Panax ginseng administration in the rat prevents myocardial ischemia-reperfusion damage induced by hyperbaric oxygen: evidence for an antioxidant intervention. *Planta Medica*, 65, 614–619.

Matsuura, H., Kasai, R., Tanaka, O., Saruwatari, Y., Kunihiro, K., and Fuwa, T. (1984) Further studies on dammarane-saponins of Ginseng root. *Chemical and Pharmaceutical Bulletin*, 32, 1188–1192.

Matsuda, H., Namba, K., Fukuda, S., Tani, T., and Kubo, M. (1986a) Pharmacological study on Panax ginseng C. A. Meyer. III. Effects of red ginseng on experimental disseminated intravascular coagulation. (2). Effects of ginsenosides on blood coagulative and fibrinolytic systems. *Chemical and Pharmaceutical Bulletin*, 34, 1153–1157.

Matsuda, H., Namba, K., Fukuda, S., Tani, T., and Kubo, M. (1986b) Pharmacological study on Panax ginseng C.A. Meyer. IV. Effects of red ginseng on experimental disseminated intravascular coagulation. (3). Effect of ginsenoside-Ro on the blood coagulative and fibrinolytic system. *Chemical and Pharmaceutical Bulletin*, **34**, 2100–2104.

Matsuzaki, Y., Matsuzaki, T., Takeda, S., Koguchi, S., Ikeya, Y., Mitsuhashi, H., Sasaki, H., Aburada, M., Hosoya, E., and Oyama, T. (1991) Studies on the metabolic fate of gomisin A (TJN-101). I. Absorption in rats. *Journal of the Pharmaceutical Society of Japan*, **111**, 524–530.

Mei, B., Wang, Y.F., Wu, J.X., and Chen, W.Z. (1994) Protective effects of ginsenosides on oxygen free radical induced damages of cultured vascular endothelial cells in vitro. *Acta Pharmaceutica Sinica*, **29**, 801–808.

Miyata, T., Fuchikamki, J., Kai, H., and Takahama, K. (1991) Characteristics of antitussive action of Mai-Men-Dong-Tang and Ophiopogonin, a steroid saponin extracted from Mai-Men-Dong. *Kanpo to Men'eki Arerugi*, **5**, 60–73.

Miyata, T., Fuchikamki, J., Kai, H., and Takahama, K. (1992) Effects of Bakumondo-to and an extracted ingredient of Bakumondo on tachykinin-induced coughing and neutral endopeptidase activity. *Kanpo to Men'eki Arerugi*, **6**, 29–37.

Miyazawa, M., Hirota, K., Fukuyama, M., Ishikawa, Y., and Kameoka, H. (1998) Insecticidal lignans against Drosophila melanogaster from fruits of Schisandra chinensis. *Natural Products Letters*, 12, 175–180.

Mizoguchi, Y., Kawada, N., Ichikawa, Y., and Tsutsui, H. (1991a) Effect of gomisin A in the prevention of acute hepatic failure induction. *Planta Medica*, **57**, 320–324.

Mizoguchi, Y., Shin, T., Kobayashi, K., and Morisawa, S. (1991b) Effect of gomisin A in an immunologically-induced acute hepatic failure model. *Planta Medica*, **57**, 11–14.

Mochizuki, M., Yoo, Y.C., Matsuzawa, K., Sato, K., Saiki, I., Tono-oka, S., Samukawa, K., and Azuma, I. (1995) Inhibitory effect of tumor metastasis in mice by saponins, ginsenoside-Rb2, 20(R)- and 20(S)-ginsenoside-Rg3, of red ginseng. *Biological and Pharmaceutical Bulletin*, **18**, 1197–1202.

Moon, J., Yu, S.J., Kim, H.S., and Sohn, J. (2000) Induction of G(1) cell cycle arrest and p27(KIP1) increase by panaxydol isolated from Panax ginseng. *Biochemical Pharmacology*, **59**, 1109–1116.

Nagai, H., Yakuo, I., Aoki, M., Teshima, K., Ono, Y., Sengoku, T., Shimazawa, T., Aburada, M., and Koda, A. (1989) The effect of gomisin A on immunologic liver injury in mice. *Planta Medica*, **55**, 13–17.

Nagai, Y., Tanaka, O., and Shibata, S. (1971) Chemical studies on the oriental plant drugs. XXIV. Structure of ginsenoside-$Rg_1$, a neutral saponin of Ginseng root. *Chemical and Pharmaceutical Bulletin*, 27, 881–892.

Niu, X.Y., Wang, W.J., Bian, Z.J., and Ren, Z.H. (1983) Effects of schizandrol on the central nervous system. *Acta Pharmaceutica Sinica*, **18**, 416–421.

Odani, T., Tanizawa, H., and Takino, Y. (1983) Studies on the absorption, distribution, excretion and metabolism of ginseng saponins. II. The absorption, distribution and excretion of ginsenoside Rg1 in the rat. *Chemical and Pharmaceutical Bulletin*, **31**, 292–298.

Odashima, S., Ohta, T., Kohno, H., Matsuda, T., Kitagawa, I., Abe, H., and Arichi, S. (1985) Control of phenotypic expression of cultured B16 melanoma cells by plant glycosides. *Cancer Research*, **45**, 2781–2784.

Ohta, Y. and Hirose, Y. (1968a) Structure of Sesquicarene. *Tetrahedron Letters*, 1251–1254.

Ohta, Y. and Hirose, Y. (1968b) New Sesquiterpenoids from *Schisandra chinensis*. *Tetrahedron Letters*, 2483–2485.

Ohtaki, Y., Nomura, M., Hida, T., Miyamoto, K., Kanitani, M., Aizawa, T., and Aburada, M. (1994) Inhibition by gomisin A, a lignan compound, of hepatocarcinogenesis by 3′-methyl-4-dimethylaminoazobenzene in rats. *Biological and Pharmaceutical Bulletin*, **17**, 808–814.

Ono, H., Matsuzaki, Y., Wakui, Y., Takeda, S., Ikeya, Y., Amagaya, S., and Maruno, M. (1995) Determination of schizandrin in human plasma by gas chromatography-mass spectrometry. *Journal of Chromatography B: Biomedical Applications*, **674**, 293–297.

Onomura, M., Tsukada, H., Fukuda, K., Hosokawa, M., Nakamura, H., Kodama, M., Ohya, M., and Seino, Y. (1999) Effects of ginseng radix on sugar absorption in the small intestine. *American Journal of Chinese Medicine*, 27, 347–354.

Oshima, Y., Konno, C., and Hikino, H. (1985) Isolation and hypoglycemic activity of panaxans I, J, K and L, glycans of Panax ginseng roots. *Journal of Ethnopharmacology*, 14, 255–259.

Pieralisi, G., Ripari, P., and Vecchiet, L. (1991) Effects of a standardized ginseng extract combined with dimethylaminoethanol bitartrate, vitamins, minerals, and trace elements on physical performance during exercise. *Clinical Therapeutics*, 13, 373–382.

Poplawski, J., Wrobel, J.T., and Glinka, T. (1980) Panaxydol, a new polyacetylenic epoxide from *Panax ginseng* roots. *Phytochemistry*, 19, 1539–1541.

Sanada, S., Kondo, N., Shoji, J., Tanaka, O., and Shibata, S. (1974b) Studies on the saponins of Ginseng. II. Structures of ginsenoside-Re, -Rf, and -$Rg_2$. *Chemical and Pharmaceutical Bulletin*, 22, 2407–2412.

Sanada, S. and Shoji, J. (1978) Studies on the saponins of Ginseng. III. Structures of ginsenoside-$Rb_3$ and 20-glucoginsenoside-Rf. *Chemical and Pharmaceutical Bulletin*, 26, 1694–1697.

Sanda, S., Kondo, N., Shoji, J., Tanaka, O., and Shibata, S. (1974a) Studies on the saponins of Ginseng. I. Structures of ginsenoside-Ro, -$Rb_1$, -$Rb_2$, -Rc and -Rd. *Chemical and Pharmaceutical Bulletin*, 22, 421–428.

Shibata, S., Tanaka, O., Shoji, J., and Saiti, H. (1985) Chemistry and pharmacology of Panax. In H. Wagner, H. Hikino, and N.R. Farnsworth (eds), *Economic and Medicinal Plant Research*, Academic Press, London, p. 217–284.

Shim, S.C., Koh, H.Y., and Han, B.H. (1983) Polyacetylenes from *Panax ginseng* roots. *Phytochemistry*, 22, 1817–1818.

Shoji, J. (1985) Recent advances in the chemical studies on ginseng. In H.M. Chang, H.W. Yeung, W.W. Tso, A. Koo, (eds), *Advances in Chinese Medicinal Material Research*, World Science, Singapore, p. 455–469.

Song, W. and Tong, Y. (1983) Occurrence and assay of some important lignans in Wu Wei Zi (*Schisandra chinensis*) and its allied species. *Acta Pharmaceutica Sinica*, 18, 138–143.

Song, W.Z., Tong, Y.Y., and Cheng, L.S. (1990) Lignans of the genus Schisandra in China. *Journal of Natural Product Research and Development*, 2, 51–58.

Sotaniemi, E.A., Haapakoski, E., and Rautio, A. (1995) Ginseng therapy in non-insulin-dependent diabetic patients. *Diabetes Care*, 18, 1373–1375.

Tada, A. and Shoji, J. (1972) Studies on the Constituents of Ophiopogonis Tube. II. On the Structure of Ophiopogonin B. *Chemical & Pharmaceutical Bulletin*, 20, 1729–1734.

Taguchi, H. and Ikeya, Y. (1975) The constituents of *Schisandra chinensis* Baill. I. The structures of gomisim A, B and C. *Chemical and Pharmaceutical Bulletin*, 23, 3296–3298.

Taguchi, H. and Ikeya, Y. (1977) The constituents of *Schisandra chinensis* Baill. The structures of two new lignans, gomisin F and G, and the absolute structures of gomisin A, B and C. *Chemical and Pharmaceutical Bulletin*, 25, 364–366.

Takahashi, M. and Yoshikura, M. (1966a) Studies on the components of *Panax ginseng* C.A. Meyer. IV. On the structure of a new acetylene derivative 'panaxynol' (2). Synthesis of 1,7,9-heptadecatrien-4-yn-3-ol. *Yakugaku Zasshi*, 86, 1051–1053.

Takahashi, M. and Yoshikura, M. (1966b) Studies on the components of *Panax ginseng* C.A. Meyer. V. On the structure of a new acetylene derivative, 'panaxynol' (3). Synthesis of 1,9-(cis)-heptadecadiene-4,6-diyn-3-ol. *Yakugaku Zasshi*, 86, 1053–1056.

Takeda, S., Kase, Y., Arai, I., Hasegawa, M., Sekiguchi, Y., Funo, S., Aburada, M., Hosoya, E., Mizoguchi, Y., and Morisawa, S. (1986) Effects of TJN-101 ((+)-(6s,7s,R-biar)-5,6,7,8-tetrahydro-1,2,3,12-tetramethoxy-6,7-dimethyl-10,11-methylenedioxy-6-dibenzo[a,c] cyclooctenol) on liver regeneration after partial hepatectomy, and on regional hepatic blood flow and fine structure of the liver in normal rats. *Folia Pharmacologica Japonica*, 88, 321–330.

Takeda, S., Arai, I., Hasegawa, M., Tatsugi, A., Aburada, M., and Hosoya, E. (1988) Effect of gomisin A (TJN-101), a lignan compound isolated from Schisandra fruits, on liver function in rats. *Folia Pharmacologica Japonica*, 91, 237–244.

Tam W. (1992) Panax ginseng C.A.Mey. In W. Tam, G. Eisenbrand (eds), *Chinese Drugs of Plant Origin : Chemistry, Pharmacology, and Use in Traditional and Modern Medicine*, Springer-Verlag, Hong Kong, p. 711–737.

Tan, R., Li, L.N., and Fang, Q.C. (1986) The stereostructure of wuweizisu B. *Planta Medica*, 52, 49–51.

Teng, C.M., Kuo, S.C., Ko, F.N., Lee, J.C., Lee, L.G., Chen, S.C., and Huang, T.F. (1989) Antiplatelet actions of panaxynol and ginsenosides isolated from ginseng. *Biochimica et Biophysica Acta*, **990**, 315–320.

Tsang, D., Yeung, H.W., Tso, W.W., and Peck, H. (1985) Ginseng saponins: influence on neurotransmitter uptake in rat brain synaptosomes. *Planta Medica*, 221–224.

Voces, J., Alvarez, A.I., Vila, L., Ferrando, A., Cabral,d. O., and Prieto, J.G. (1999) Effects of administration of the standardized Panax ginseng extract G115 on hepatic antioxidant function after exhaustive exercise. *Comparative Biochemistry and Physiology, Part C Pharmacology, Toxicology, Endocrinology*. **123**, 175–184.

Wakabayashi, C., Hasegawa, H., Murata, J., and Saiki, I. (1997) In vivo antimetastatic action of ginseng protopanaxadiol saponins is based on their intestinal bacterial metabolites after oral administration. *Oncology Research*, **9**, 411–417.

Wakabayashi, C., Murakami, K., Hasegawa, H., Murata, J., and Saiki, I. (1998) An intestinal bacterial metabolite of ginseng protopanaxadiol saponins has the ability to induce apoptosis in tumor cells. *Biochemical and Biophysical Research Communications*, **246**, 725–730.

Wang, B.X., Yang, M., Jin, Y.L., Cui, Z.Y., and Wang, Y. (1990) Studies on the hypoglycemic effect of ginseng polypeptide. *Acta Pharmaceutica Sinica*, **25**, 401–405.

Wen, T.C., Yoshimura, H., Matsuda, S., Lim, J.H., and Sakanaka, M. (1996) Ginseng root prevents learning disability and neuronal loss in gerbils with 5-minute forebrain ischemia. *Acta Neuropathologica*, **91**, 15–22.

Xiaoguang, C., Hongyan, L., Xiaohong, L., Zhaodi, F., Yan, L., Lihua, T., and Rui, H. (1998) Cancer chemopreventive and therapeutic activities of red ginseng. *Journal of Ethnopharmacology*, **60**, 71–78.

Xu, S.X., Wang, Q.H., Zhang, G.G., Lu, Y.J., Chen, G.X., and Chen, Y.J. (1988) Studies on the Chemical Constituents of *Panax ginseng* C.A. Meyer XII. Study on the Liposoluble Constituents of *Panax ginseng* C.A. Meyer. *Journal of Shenyang College of Pharmacy*, **5**, 16–19.

Yahara, S., Sakamoto, C., Nohara, T., Niiho, Y., Nakajima, Y., and Ito, H. (1993) Thymoquinol Glucosides from Schisandrae Fructus. *Japanese Journal of Pharmacognosy*, **47**, 420–422.

Yamada, S., Murawaki, Y., and Kawasaki, H. (1993) Preventive effect of gomisin A, a lignan component of shizandra fruits, on acetaminophen-induced hepatotoxicity in rats. *Biochemical Pharmacology*, **46**, 1081–1085.

Yamaguchi, Y., Haruta, K., and Kobayashi, H. (1995) Effects of ginsenosides on impaired performance induced in the rat by scopolamine in a radial-arm maze. *Psychoneuroendocrinology*, **20**, 645–653.

Yamamoto, M., Masaka, M., Yamada, K., Hayashi, Y., Hirai, A., and Kumagai, A. (1978) Stimulatory effect of ginsenosides on DNA, protein and lipid synthesis in rat bone marrow and participation of cyclic nucleotides. *Arzneimittel-Forschung*, **28**, 2238–2241.

Yang, M., Wang, B.X., Jin, Y.L., Wang, Y., and Cui, Z.Y. (1990) Effects of ginseng polysaccharides on reducing blood glucose and liver glycogen. *Acta Pharmacologica Sinica*, **11**, 520–524.

Yang, Y., Wu, T., He, K., and Fu, Z.G. (1999) Effect of aerobic exercise and ginsenosides on lipid metabolism in diet-induced hyperlipidemia mice. *Acta Pharmacologica Sinica*, **20**, 563–565.

Yang, Z., Xiao, R., and Xiao, Z.Y. (1987a) Study on Chemical Constituents of *Ophiopogon japonicus* (Thumb) of Sichuan (I). *West China Journal of Pharmaceutical Science*, **2**, 57–60.

Yang, Z., Xiao, R., and Xiao, Z.Y. (1987b) Study on Chemical Constituents of *Ophiopogon japonicus* (Thumb) of Sichuan (II). *West China Journal of Pharmaceutical Science*, **2**, 121–124.

Yasukawa, K., Ikeya, Y., Mitsuhashi, H., Iwasaki, M., Aburada, M., Nakagawa, S., Takeuchi, M., and Takido, M. (1992) Gomisin A inhibits tumor promotion by 12-O-tetradecanoylphorbol-13-acetate in two-stage carcinogenesis in mouse skin. *Oncology*, 49, 68–71.

Yim, T.K. and Ko, K.M. (1999) Schisandrin B protects against myocardial ischemia-reperfusion injury by enhancing myocardial glutathione antioxidant status. *Molecular and Cellular Biochemistry*, 196, 151–156.

Yu, B.Y., Yin X., Zhang, C.H., and Xu, G.J. (1991) Immunity study on polysaccharide from tuberous roots of Ophiopogon japonicus (L. f.) Ker-Gawl. *Journal of Chinese Pharmaceutical University*, 22, 286–288.

Yun, T.K. (1996) Experimental and epidemiological evidence of the cancer-preventive effects of Panax ginseng C.A. Meyer. *Nutrition Reviews*, 54, S71–S81.

Yun, T.K. (1999) Update from Asia. Asian studies on cancer chemoprevention. *Annals of the New York Academy of Sciences*, 889, 157–192.

Yun, Y.S., Moon, H.S., Oh, Y.R., Jo, S.K., Kim, Y.J., and Yun, T.K. (1987) Effect of red ginseng on natural killer cell activity in mice with lung adenoma induced by urethan and benzo(a)pyrene. *Cancer Detection and Prevention*, **Supplement.** 1, 301–309.

Zhang, G. and Ding, W. (1962) A preliminary report on the phytophthora root rot of Panax ginseng C.A. Meyer. *China Journal of Chinese Materia Medica*, 15, 18–19.

Zhang, H.X., Sun, Y.X., Wang, S.Q., Jiang, W.P., and Yang, L.R. (1985) Analysis of the volatile constituents of Jilin ginseng. *Chinese Science Bulletin*, 30, 195–199.

Zhang, L.L., Xiao, Y.L., and Li, X.G. (1989) Isolation and Identification of the Salicymide of Red Ginseng. *Zhong Yao Cai*, 12, 33–34.

Zhang, T.M., Wang, B.E., and Liu, G.T. (1989) Action of schizandrin B, an antioxidant, on lipid peroxidation in primary cultured hepatocytes. *Acta Pharmacologica Sinica*, 10, 353–356.

Zhang, T.M., Wang, B.E., and Liu, G.T. (1992) Effect of schisandrin B on lipoperoxidative damage to plasma membrane of rat liver in vitro. *Acta Pharmacologica Sinica*, 13, 255–258.

Zhang, W.X. and Wang, N.H. (1993) Hypoglycemic action of the polysaccharides from Radix Ophiopogonis in alloxan-induced diabetic mice. *Chinese Traditional and Herbal Drugs*, 24, 30–31.

Zhu, Y.P. (1998a) Mai Dong. In Y.P. Zhu (ed.), *Chinese Materia Medica: Chemistry, Pharmacology and Applications*, Harwood Academic Publishers, Amsterdam, p. 629–632.

Zhu, Y.P. (1998b) Wu Wei Zi. In Y.P. Zhu (ed.), *Chinese Materia Medica: Chemistry, Pharmacology and Applications*, Harwood Academic Publishers, Amsterdam, p. 653–657.

Zhu, Y.X., Yan, D.K., and Tu, G.S. (1987) Isolation and Identification of Homoisoflavanones from Maidong (*Ophiopogon japonicus* (Thumb) Ker-Gawl). *Acta Pharmaceutica Sinica*, 22, 679–684.

Zhu, Y.X., Liu, L.Z., Lin, D.K., and Wang, W. (1989) Studies on Chemical Constituents of *Ophiopogon japonicus*. *China Journals of Chinese Materia Medica*, 14, 359–360.

Zhu, Y.X., Liu, L.Z., Wang, W., Ling, D.K., and Sun, Z.P. (1991) Studies on Chemical Constituents of Essential Oil of *Ophiopogon japonicus*. *Chinese Journal of Pharmaceutical Analysis*, 11, 21–23.

Ziemba, A.W., Chmura, J., Kaciuba-Uscilko, H., Nazar, K., Wisnik, P., and Gawronski, W. (1999) Ginseng treatment improves psychomotor performance at rest and during graded exercise in young athletes. *International Journal of Sport Nutrition*, 9, 371–377.

# 6 Contemporary applications of a traditional formula: manufacture of Shengmai San (SMS) preparations and their applications

*Liang-Yuan Zheng*
[Translated by Kam-Ming Ko]

Shengmai San (SMS), also known as Shengmai Yin, is a traditional Chinese medicinal formula that is usually administered as a decoction. Over the past decades, a number of modern dosage forms have evolved from the modification of the original formula including the *Codonopsis*- and *Ginseng*-formula (Chen 1998; Hu *et al.* 1998). Although the official dosage form of SMS described in the Year 2000 edition of the Chinese Pharmacopoeia (CP) is an oral tonic, official standards of other dosage forms, such as capsule, syrup, granule, tea bag, tablet, and injection have also been established by the Ministry of Health, China.

## RECENTLY STANDARDIZED FORMULATIONS AND DOSAGE

### I. Year 2000 edition of the Chinese Pharmacopoeia

#### *A. Shengmai tonic*

Formulation (Ministry of Health, China 2000): *Radix Ginseng* 100 g, *Radix Ophiopogonis* 200 g, and *Fructus Schisandrae* 100 g for every 1000 ml of extract. Taken thrice daily, 10 ml each time, equivalent to a daily consumption of *Radix Ginseng* 3 g, *Radix Ophiopogonis* 6 g, and *Fructus Schisandrae* 3 g.

## RECOGNIZED STANDARD PREPARATIONS OF SMS BY THE MINISTRY OF HEALTH, CHINA

### I. Shengmai capsules

Formulation (Ministry of Health, China 1993): *Radix Ginseng* 330 g, *Radix Ophiopogonis* 660 g, and *Fructus Schisandrae* 330 g for every 1000 capsules. Taken thrice daily, 3 capsules each time, equivalent to a daily consumption of *Radix Ginseng* 2.97 g, *Radix Ophiopogonis* 5.94 g, and *Fructus Schisandrae* 2.97 g.

## II. Shengmai tonic

Formulation (Ministry of Health, China 1995): *Radix Codonopsis* 300 g, *Radix Ophiopogonis* 200 g, and *Fructus Schisandrae* 100 g for every 1000 ml of extract. Taken thrice daily, 10 ml each time, equivalent to a daily consumption of *Radix Codonopsis* 9 g, *Radix Ophiopogonis* 6 g, and *Fructus Schisandrae* 3 g.

## III. Shengmai syrup (Codonopsis-formula)

Formulation (Ministry of Health, China 1996a): same as Shengmai Tonic.

## IV. Shengmai granules (Codonopsis-formula)

Formulation (Ministry of Health, China 1996b): *Radix Codonopsis* 300 g, *Radix Ophiopogonis* 200 g, *Fructus Schisandrae* 100 g, and excipients for granulation. Taken thrice daily, 10 g each time.

## V. Shengmai tea-bag

Formulation (Ministry of Health, China 1996c): *Radix Ginseng* 100 g, *Radix Ophiopogonis* 200 g, and *Fructus Schisandrae* 100 g for every 1000 tea-bags. Taken thrice daily, 0.4 g (1 tea-bag) each time by soaking in hot water. The dosage is equivalent to a daily consumption of *Radix Ginseng* 0.3 g, *Radix Ophiopogonis* 0.6 g, and *Fructus Schisandrae* 0.3 g.

## VI. Shengmai tablets (Codonopsis-formula)

Formulation (Ministry of Health, China 1997a): *Radix Codonopsis* 300 g, *Radix Ophiopogonis* 200 g, and *Fructus Schisandrae* 100 g for every 1000 tablets. Taken thrice daily, 8 tablets each time, equivalent to a daily consumption of *Radix Codonopsis* 7.2 g, *Radix Ophiopogonis* 4.8 g, and *Fructus Schisandrae* 2.4 g.

## VII. Shengmai injection

Formulation (Ministry of Health, China 1997b): *Radix Ginseng Destillata* 100 g, *Radix Ophiopogonis* 312 g, and *Fructus Schisandrae* 156 g for every 1000 ml; standard preparations of 2 ml, 10 ml and 20 ml are available. Administered 1–2 times a day, by intramuscular injection at a dose of 2–4 ml each time, equivalent to a daily consumption of *Radix Ginseng Destillata* 0.4–0.8 g, *Radix Ophiopogonis* 1.25–2.5 g, and *Fructus Schisandrae* 0.62–1.25 g. Intravenous infusion: 20–60 ml each time, equivalent to a daily consumption of *Radix Ginseng Destillata* 2–6 g, *Radix Ophiopogonis* 6.2–18.7 g, and *Fructus Schisandrae* 3.12–9.36 g.

## CONTEMPORARY DOSAGE FORMS AND PROCESSING TECHNIQUES

### I. Shengmai tonic (oral liquid)

#### *A. Preparation*

The three component herbs, *Radix Ginseng*, *Radix Ophiopogonis*, and *Fructus Schisandrae* are ground into powder and processed as described in the Chinese Pharmacopoeia (under the Percolation section in Appendix 10: Extract and Liquid Extraction). Using 65 per cent ethanol as the solvent, the herb powder is soaked for 24 hours before percolation. A total of 4500 ml of percolate is collected and concentrated to 250 ml at reduced pressure. After cooling, the concentrate is diluted with 400 ml of water, filtered, and mixed with 300 ml of 60 per cent syrup and a suitable amount of preservative. The pH value of the mixture is then adjusted to a suitable range and made up to 1000 ml, allowed to settle, filtered, bottled, and then sterilized (Ministry of Health, China 2000).

#### *B. Process improvement studies:*

(a) The degree of translucence of the mixture could be enhanced by de-fatting with paraffin after the adjustment of pH. (b) The degree of translucence of the mixture could also be enhanced by adjusting the pH value to 8, the intermediate was allowed to settle before diluting to oral liquid. (c) The effect of fine processing using the natural flocculant, chitosan, has been found to be superior to ultracentrifugation and as good as alcohol precipitation (Zhang 1994; Chang *et al.* 1998; Shao 1999).

### II. Shengmai capsules

#### *A. Preparation*

Two hundred grams of the *Ginseng* is ground into fine powder and left aside for use in later processes. The rest of the *Ginseng* (130 g) is ground into coarse powder, soaked in 75 per cent ethanol for 24 hours, and extracted by percolation as described in the Appendix (p.17) in the Chinese Pharmacopoeia. *Fructus Schisandrae* is ground into coarse powder and extracted by distillation. The residue, together with *Radix Ophiopogonis*, is extracted twice with boiling water, filtered, and combined. The combined extract is then concentrated, made up with ethanol to contain 60 per cent (v/v) of ethanol in composition, allowed to settle and filtered. The filtrate is then concentrated (with the ethanol recycled) and combined with the *Ginseng* percolate, *Schisandrae* distillate (volatile oil) and the *Ginseng* fine powder. The mixture is then granulated, dried, and filled into 1000 capsules (Ministry of Health, China 1993).

### III. Shengmai tonic (Codonopsis-formula)

#### *A. Preparation*

The 3 component herbs, *Radix Codonopsis*, *Radix Ophiopogonis*, and *Fructus Schisandrae*, are extracted twice with boiling water: 2 hours for the first extraction and 1.5 hours for the second. The extracts are combined, filtered, and concentrated to a volume of about 300 ml and cooled. Ethanol (600 ml) is then added, allowed to stand for 24 hours and filtered. The filtrate is then concentrated at reduced pressure to form a paste, after which a trace amount of water is added to facilitate filtration. Mono-sugar syrup (300 ml) and preservative are added to this second filtrate, and made up to 1000 ml with water. The mixture is allowed to settle and then filtered. The product is ready after bottling and sterilization (Ministry of Health, China 1995).

### *B. Studies on ultrafiltration technique*

In the fine processing of Shengmai tonic, ultrafiltration using external pressure and hollow fiber offers the benefits of short production cycle and high cost efficiency. Accelerated studies with ultrafiltration technique showed that it could raise the composition of the active ingredients, without compromising translucence and hence the degree of precipitation (Liu *et al.* 1996).

## IV. Shengmai syrup (*Codonopsis*-formula)

### *A. Preparation*

The three component herbs, *Radix Codonopsis*, *Radix Ophiopogonis* and *Fructus Schisandrae*, are extracted twice with boiling water; 2 hours for the first extraction and 0.5 hours for the second. The extracts are combined, filtered and concentrated to a suitable volume, after which ethanol is added to make up a final ethanol composition of 60 per cent and allowed to settle for 24 hours. The mixture is then filtered, and the filtrate is concentrated at reduced pressure. Cane sugar (650 g) is added to the concentrate with stirring until it is dissolved. The syrup is then boiled for 30 minutes, after which preservative is added and made up to 1000 ml with water (Ministry of Health, China 1996a).

## V. Shengmai granules (*Codonopsis*-formula)

### *A. Preparation*

The three component herbs, *Radix Codonopsis*, *Radix Ophiopogonis* and *Fructus Schisandrae*, are extracted twice with boiling water: 2 hours for the first extraction and 1.5 hours for the second. The extracts are combined, filtered, and concentrated to a volume of about 300 ml and cooled. Ethanol (600 ml) is then added, allowed to stand for 24 hours and filtered. The filtrate is then concentrated at reduced pressure until it reaches a relative density of 1.30(60–65°C) and appears as a clear paste. The paste is mixed with five parts of cane sugar, granulated and dried (Ministry of Health, China 1996b).

## VI. Shengmai tea-bag

### *A. Preparation*

The three component herbs, *Radix Codonopsis*, *Radix Ophiopogonis* and *Fructus Schisandrae* are ground into coarse powder, mixed, and granulated. The granules could then be filled into 1000 tea-bags after drying (Ministry of Health, China 1996c).

## VII. Shengmai tablets (*Codonopsis*-formula)

### *A. Preparation*

A portion of the *Radix Codonopsis* is ground into powder, and the fine powder (about 240 g) is reserved for future use. The volatile oil of *Fructus Schisandrae* is extracted by distillation. The aqueous distillant is collected in a separate container, while the residue together with the coarse powder of *Radix Codonopsis* and *Radix Ophiopogonis*, are extracted twice with water and then filtered. The extract is combined with the aqueous distillant of *Fructus Schisandrae*, concentrated to paste form, and mixed with the fine powder of *Radix Codonopsis* and excipients. After granulation, the granules are dried and sprayed with the volatile oil of *Fructus Schisandrae*. The process is completed by pressing the granules into 1000 tablets and coating with sugar (Ministry of Health, China 1997a).

## VIII. Shengmai injection

### *A. Preparation*

*Radix Ginseng Destillata* is ground into fine granules, and then extracted 4–5 times with ethanol (each for 2 hours). The end-point of extraction could be monitored by thin-layer chromatography (TLC). The extracts are combined, chilled, filtered, and concentrated to paste form, and mixed with 400 ml injection grade water. The mixture is then chilled, filtered, and reserved for future use. *Fructus Schisandrae* is distilled with steam until 150 ml of distillate is collected, which is refrigerated for later use. The residue is extracted three times with water, 40 minutes each time, and the extracts are combined, filtered, and concentrated to paste form. The paste is then subjected to ethanol precipitation twice, at final ethanol concentrations of 80 and 85 per cent respectively. The ethanol solutions are then filtered, with the filtrate being combined and ethanol being recycled in subsequent concentration process. The filtrate is then concentrated to form a paste. The paste is then mixed with 200 ml injection grade water, chilled, and filtered, after which the filtrate is boiled for 30 minutes with activated charcoal. The solution is cooled for a while, and then filtered for later use. A clear aqueous extract of *Radix Ophiopogonis* (about 200 ml) is prepared in the same way of *Fructus Schisandrae*. In preparing the final dosage form, the aqueous extracts of *Radix Ginseng Destillata*, *Fructus Schisandrae*, and *Radix Ophiopogonis* and the distillate of *Fructus Schisandrae* are mixed, filtered and made up to 1000 ml with injection grade water. The pH of the solution is then adjusted to 7.5, filtered, bottled, sterilized, and ready for use (Ministry of Health, China 1997b).

# QUALITY CONTROL AND STANDARDIZATION

## I. Shengmai tonic (oral liquid)

### *A. Physical attributes*

A clear liquid that is brownish yellow to pale red in color; mild turbidity on prolonged standing; fragrant in scent; sour, sweet, and lightly bitter in taste (Ministry of Health, China 2000).

### *B. Authentication*

(a) An aliquot of the sample (20 ml) is partitioned with 20 ml n-butanol. The butanol fraction is evaporated to dryness, and the residue is refluxed with 15 ml sulfuric acid-acidified ethanol (45 per cent)(7:100 v/v) for 1 hour. The ethanol is evaporated, and the aqueous solution is partitioned with 10 ml of chloroform. The chloroform fraction is taken and dehydrated with anhydrous sodium sulfate. After filtration, the solvent is concentrated to a final volume of 1 ml, serving as the test sample. A mixed-standard solution containing 1 mg/ml each of panaxadiol and panaxatriol is prepared by dissolving the standards in anhydrous ethanol. Qualitative analysis is performed using the thin-layer chromatographic (TLC) method as described in Appendix VI B of the Chinese Pharmacopoeia. Aliquots (10 μl) of the test sample and standard solution are applied to a silica-G plate, and eluted with cyclohexane-propanol (2:1). After drying, the TLC plate is sprayed with sulfuric acid-methanol (1:1) solution, heated at 105°C for 10 minutes, and observed under UV light (365 nm). The test sample should have fluorescent spots appearing at the equivalent positions as compared with standards.

(b) An aliquot of the sample (10 ml) is mixed with 0.5 ml hydrochloric acid and 1 ml water. The mixture is boiled for 5 minutes, allowed to cool down, and then partitioned with 20 ml of chloroform. The chloroform fraction is concentrated to a final volume of 1 ml, serving as the test sample. A standard herb of *Radix Ophiopogonis* (1 g) is heated with 20 ml of water for 10 minutes. The extract could then serve as the standard solution after filtration and addition of 0.5 ml hydrochloric acid. Analysis is performed using TLC method as described in Appendix VI B of the Chinese Pharmacopoeia. Aliquots (5 μl) of the test sample and standard solution are applied to a silica-G plate, and eluted with chloroform-propanol (4:1). After drying, the TLC plate is sprayed with 10 per cent sulfuric acid-ethanol solution, and heated at 100°C until the color spots become clear and vivid. The test sample should have color spots appearing at the equivalent positions as compared with standards.

### *C. Physicochemical analyses*

Relative density not less than 1.08 (Appendix VII A, CP). pH value should fall between 4.5 and 7.0 (Appendix VII G, CP). Others should meet the relevant requirements for mixture in the Chinese Pharmacopoeia (Appendix I,J).

## II. Shengmai capsules

### *A. Physical attributes*

The product exists in capsule form, with brownish-yellow powdery content; fragrant in scent; sour, sweet, and lightly bitter in taste (Ministry of Health, China 1993).

### *B. Authentication*

The content of 10 capsules is dissolved in 20 ml of water, put into a separating funnel, and partitioned with n-butanol (20 ml). The butanol fraction is then filtered and evaporated to dryness. The residue is then reconstituted with 15 ml of a 7 per cent sulfuric acid-45 per cent ethanol (7.4:93) solution, and refluxed for 1 hour, after which the ethanol component is evaporated, and the remaining solvent is partitioned with 10 ml of chloroform. The chloroform fraction is taken out, dehydrated with anhydrous sodium sulfate, filtered, and concentrated to 1 ml, serving as the test sample. A mixed-standard solution containing 1 mg/ml each of panaxadiol and panaxatriol is prepared by dissolving the standards in anhydrous ethanol. Qualitative analysis is performed using TLC method as described in Appendix VI B of the Chinese Pharmacopoeia. Aliquots (10 μl) of the test sample and standard solution are applied to a silica-G plate, and eluted with cyclohexane-propanol (2:1). After drying, the TLC plate is sprayed with sulfuric acid-methanol (1:1) solution, heated at 105°C for 10 minutes, and observed under UV light (365 nm). The test sample should have fluorescent spots appearing at the equivalent positions as the standards.

### *C. Physicochemical analyses*

Moisture content is determined by Method 1 as described in p.30 of the Appendix (C.P.) and should not exceed 7.5 per cent. Others should meet the relevant requirements for capsules in the Chinese Pharmacopoeia (Appendix p.16).

## III. Shengmai tonic (*Codonopsis*-formula)

### *A. Physical attributes*

The product exists as a brownish-yellow liquid; fragrant in scent; sweet and sour in taste (Ministry of Health, China 1995).

### *B. Physicochemical analyses*

Relative density not less than 1.08 (C.P. Appendix p.34). Others should meet the relevant requirements for mixture in the Chinese Pharmacopoeia (Appendix p.15).

## IV. Shengmai syrup (*Codonopsis*-formula)

### *A. Physical attributes*

The product exists as a brownish paste; fragrant in scent; sweet in taste (Ministry of Health, China 1996a).

### *B. Physicochemical analyses*

Relative density not less than 1.24 (C.P. Appendix VII A). Others should meet the relevant requirements for syrup in the Chinese Pharmacopoeia (Appendix I H).

## V. Shengmai granules (*Codonopsis*-formula)

### *A. Physical attributes*

The product exists as brownish-yellow granules; fragrant in scent; sweet in taste (Ministry of Health, China 1996b).

### *B. Physicochemical analyses*

It should meet the relevant requirements for granules in the Chinese Pharmacopoeia (Appendix I C).

## VI. Shengmai tea-bag

### *A. Physical attributes*

The product exists as brownish-yellow granules; fragrant in scent; astringent and slightly bitter in taste (Ministry of Health, China 1996c).

### *B. Authentication*

(a) Microscopic examination reveals resin canals that contain yellowish secretory droplets or granules; round cluster crystals of calcium oxalate with a diameter of 20–68 μm and sharp edges; needle-like crystals of calcium oxalate, 24–50 μm in length and 3 μm in diameter, exist in a clustered or dispersed pattern. Pale brownish-yellow sclereid (stone cells) in the seed-coat and epidermis, characterized by their polygonal appearance, thick walls, fine pores, and pale brownish substances in the cellular space.

(b) Sample of the product (4 g) is soaked in boiling water (50 ml) for 10 minutes, with gentle heating to maintain mild boiling. The soaking solution is then decanted, cooled, and partitioned with 50 ml n-butanol. The butanol fraction is then evaporated to dryness. The residue is reconstituted with 15 ml of a sulfuric acid-45 per cent ethanol (7:10) solution, and refluxed for 1 hour, after which the ethanol component is evaporated (about 7 ml solvent will remain) and made up to 10 ml with water. The solution is partitioned with 20 ml chloroform, which is then taken out, dehydrated with anhydrous sodium sulfate, filtered, and concentrated to 1 ml, serving as the test sample. A mixed-standard solution containing 1 mg/ml each of panaxadiol and panaxatriol is prepared by dissolving the standards in anhydrous ethanol. Qualitative analysis is performed using the TLC method as described in Appendix VI B of the Chinese Pharmacopoeia. Aliquots (6 μl) of the test sample and standard solution are applied to a silica-G plate and eluted with cyclohexane-propanol (2:1). After drying, the TLC plate is sprayed with a 10 per cent sulfuric acid-methanol solution, heated at 105°C for 10 minutes, cooled, and observed under UV light (365 nm). The test sample should have fluorescent spots appearing at the equivalent positions as compared with standards.

### *C. Physicochemical analyses*

The water soluble content of the product is determined by the heat soaking extraction method as described in the Chinese Pharmacopoeia (Appendix X A), except that the solution is kept boiling for 10 minutes. The total soluble content should not be less than 50.0 per cent. Others should meet the relevant requirements for tea preparations in the Chinese Pharmacopoeia (Appendix I T).

## VII. Shengmai tablets (*Codonopsis*-formula)

### *A. Physical attributes*

The product exists as a coated tablet with brownish content after the removal of the coating; sweet and sour in taste.

### *B. Authentication*

(a) Sample of the product (three tablets) is de-coated, ground into powder, and suspended in 15 ml of 50 per cent ethanol. After filtration, three drops of the filtrate is spotted onto a filter paper, dried in air, and sprayed with 0.1 per cent bromophenol blue-ethanol solution. Yellow spots should appear in the blue background. When two drops of 0.2 per cent bromcresol green-ethanol solution are added to 2 ml of the filtrate, the solution turns yellow-green following the addition of a drop of 10 per cent sodium hydroxide.

(b) A sample of the product (5 g) is refluxed with 30 ml n-butanol for 2 hours, filtered, and the filtrate is concentrated to 2 ml, serving as the test sample. Standard solution of *Radix Codonopsis* is prepared by refluxing 5 g of the herb with 30 ml n-butanol as described for the sample. Analysis is performed using the TLC method as described in Appendix VI B of the Chinese Pharmacopoeia. Aliquots (1 μl) of the test sample and standard solution are applied to a silica-G plate, and eluted with n-butanol-ethanol-water (25:3:2). After drying, the TLC plate is stained by spraying it with a 10 per cent sulfuric acid-ethanol solution and then heating at 105°C for 10 minutes. The test sample should have grayish brown spots appearing at the equivalent positions as compared with standards.

### *C. Physicochemical analyses*

Others should meet the relevant requirements for tablets in the Chinese Pharmacopoeia (Appendix I D).

## VIII. Shengmai injection

### *A. Physical attributes*

The product exists as pale yellow or brownish-yellow, clear liquid (Ministry of Health, China 1997).

### *B. Authentication*

(a) A sample of the product (10 ml) is evaporated to dryness in a water bath, and the residue is dissolved in 2 ml of ethanol, serving as the test sample. A mixed-standard solution containing 2 mg/ml each of ginsenosides $Rb_1$, Re and $Rg_1$ are prepared in ethanol. Qualitative analysis is performed using the TLC method as described in Appendix VI B of the Chinese Pharmacopoeia. Aliquots (2–4 μl) of the test sample and standard solution are applied to a silica-G plate, and eluted with chloroform-methanol-water (75:20:2). After drying, the TLC plate is stained by spraying it with a 10 per cent sulfuric acid-ethanol solution and then heating at 105°C for a few minutes. After cooling, the TLC plate is observed under UV light (365 nm). The test sample should have fluorescent spots appearing at the equivalent positions as compared with standards.

(b) Sample of the product (40 ml) is mixed with 3 ml hydrochloric acid, and heated in a water bath for 1 hour. After cooling, the solution is partitioned with 30 ml diethyl ether. The ethereal layer is dried, and the residue is dissolved in 1ml chloroform, serving as the test sample. Authenticated *Radix Ophiopogonis* (2 g) is extracted by heating with water for 30 minutes, being filtered and concentrated to about 40 ml. A standard solution is then prepared as described for test samples. Qualitative analysis is performed using TLC method as described in Appendix VI B of the Chinese Pharmacopoeia. Aliquots (5–10 μl) of the test sample and standard solution are applied to a silica-G plate, and eluted with chloroform-propanol (4:1). After drying, the TLC plate is stained by spraying it with a 10 per cent sulfuric acid-ethanol solution, and then being heated at 105°C for 5 minutes. The test sample should have colored spots appearing at the equivalent positions as compared with standards.

(c) Sample of the product (50 ml) is concentrated to 25 ml in a water bath and transferred to a separating funnel. It is then partitioned three times with chloroform (10 ml each time), filtered, evaporated to near dryness, and then reconstituted with 1 ml of chloroform, serving as the test sample. Standard solution of schisahenol (0.5 mg/ml) is prepared in chloroform. Qualitative analysis is performed using the TLC method as described in Appendix VI B of the Chinese Pharmacopoeia. Aliquots (5–10 μl) of the test sample and standard solution are applied to a silica-$GF_{254}$ plate, and eluted with petroleum ether (30–60°C)-chloroformate-formic acid (14:5:1). After drying, the TLC plate is observed under UV light (254 nm). The test sample should have colored spots appearing at the equivalent positions as compared with standards.

### *C. Physicochemical analyses*

The pH value should be 5.0–7.0. A microbial test should meet the requirements described by the Chinese Pharmacopoeia (Appendix VIII B). A pyrogen test should be performed at a dose of 2ml/kg in a rabbit and meet the requirements as described (C.P. Appendix VIII B).

Hemolytic test (Ham's Test)

(a) Preparation of 2 per cent RBC suspension: Blood obtained from a rabbit by cardiac puncture is collected into a glass bead-containing vessel. Fibrinogen is removed by sonicating the blood sample for a few minutes. The defibrinated blood is mixed with physiological saline, centrifuged, and the clear supernatant is decanted. The sedimented red blood cells are washed 3–4 times with physiological saline until the supernatant does not turn red after centrifugation. A 2 per cent (v/v) RBC suspension is prepared by reconstituting the red cells in physiological saline and used on the same day.

(b) Test Method: with five sterile test tubes, 0.3 ml of the test sample and 2.2 ml physiological saline are added to three of them, while saline (2.5 ml) or distilled water (2.5 ml) is added to the other two, serving as the negative and positive controls respectively. The RBC suspension (2 per cent) (2.5 ml) is then added to each of the five test tubes, mixed, and incubated at 36.5 ± 0.5°C. There should be no signs of hemolysis within 3 hours.

Others should meet the relevant requirements for injection preparations in the Chinese Pharmacopoeia (Appendix I U).

### *D. Quantitation of ingredients: HPLC method (C.P. Appendix VI D)*

Chromatographic conditions and system compatibility: $C_{18}$-silica column; acetonitrile-0.05 per cent phosphoric acid (19:81) as the mobile phase; detection wavelength at 203 nm. The number of theoretic plates for ginsenoside Re should not be less than 4000.

Preparation of standard solutions: Ginsenoside $Rg_1$ (15 mg) and ginsenoside Re (12 mg), having been demoisturized to stable weight in a $P_2O_5$-containing dessicator, are dissolved and made up to 25 ml with acetonitrile-water (19:81) in a volumetric flask. The solution is then diluted two-fold with the same solvent in a 10 ml volumetric flask, making final concentrations of 0.3 mg/ml $Rg_1$ and 0.24 mg/ml Re.

Test sample preparation: Sample of the product (10 ml) is evaporated to near dryness, and reconstituted in 2 ml of the mobile phase.

Quantitation: 20 μl of the test sample or standard solution is injected into the HPLC system and quantitated accordingly.

Standards: The product should contain not less than 0.08 mg/ml ginsenoside $Rg_1$ or 0.04 mg/ml ginsenoside Re.

## CONTEMPORARY RESEARCH IN QUALITY CONTROL

### I. Authentication

(a) The content of 10 capsules is dissolved in 20 ml water, transferred to a separating funnel, partitioned with 20 ml n-butanol, allowed to settle and then filtered. The butanol extract is evaporated to dryness, reconstituted, and refluxed with 15 ml of sulfuric acid-45 per cent ethanol (7.4:93) solution. The ethanol component is evaporated, and the aqueous solution is partitioned with 10 ml of chloroform. The chloroform fraction is collected and dehydrated with anhydrous sodium sulfate. After filtration, the solvent is concentrated to a final volume of 1 ml, serving as the test sample. A mixed-standard solution containing 1 mg/ml each of panaxadiol and panaxatriol is prepared by dissolving the standards in anhydrous ethanol. Qualitative analysis is performed using the TLC method. Aliquots (10 μl) of the test sample and standard solution are applied to a silica-G plate, and eluted with cyclohexane-propanol (2:1). After drying, the TLC plate is stained with sulfuric acid-methanol (1:1) solution, heated at 105°C for 10 minutes, and observed under UV light (365 nm). The test sample should have fluorescent spots appearing at the equivalent positions as compared with standards (Yan *et al.* 1990; Xie 1990).

(b) Sample of tonic or injection (1 ml) is evaporated to dryness over a water bath, and reconstituted in 1 ml ethanol, serving as the test sample. Qualitative analysis is performed using the TLC method. Aliquots (3–5 μl) of the test sample and standard solution are applied to a silica-G plate, and eluted with chloroform-methanol-water (80:20:2) for about 15 cm. After drying, the TLC plate is stained by spraying it with a 10 per cent phosphomolybdenic acid-ethanol solution and heating at 105°C for a few minutes. The chromatogram should show not less than twelve blue spots (Yan *et al.* 1990; Xie 1990).

(c) Using tonic or injection as the test samples and extracts of *Fructus Schisandrae*, *Radix Ginseng*, and *Radix Ophiopogonis* as standards, qualitative analysis is performed using the TLC method (Zhe *et al.* 1988a). Aliquots (10 μl) of the test sample and standard solution are applied to a silica-G plate, and eluted with chloroform-methanol-water (80:20:2). The TLC plate is stained with iodine vapor or observed under UV light (254 nm). The test sample should have fluorescent or colored spots appearing at the equivalent positions as compared with standards (Yan *et al.* 1990; Xie 1990).

## II. Quantitation of active ingredients

(a) Total saponin content of *Ginseng* (Zhu *et al.* 1989): (i) Preparation of test sample solutions – 10 ml water-saturated n-butanol is added to five separating funnels, numbered 1 through to 5. Shengmai tonic (10 ml) is added to the first funnel, and the counter-current extraction is performed. Each time the aqueous layer is transferred to the subsequent funnel, the butanol fraction is extracted with 10 ml fresh distilled water, yielding a total of five aqueous extracts and five butanol extracts. The aqueous extracts are combined, and extracted three times with 10 ml water-saturated n-butanol. All butanol extracts are combined and the solvent was recycled under reduced pressure. The residue is reconstituted with methanol, transferred to a 5 ml vessel, and made up to the mark. (ii) TLC Analysis – test samples (20–40 μl) containing about 100 μl *Ginseng* saponins are applied to a silica-G TLC plate. After the spots are dried, the plate is eluted with chloroform-methanol-water (70:55:10) for 15 cm. The elution tank should have been saturated with glacial acetic acid, placed in a small beaker, for 15 minutes before the analysis. The plate is stained by placing the plate in an iodine vapor-saturated tank for 1 minute, and the positions of the ginsenosides are marked. (iii) Colorimetric assay: spots on the silica-G plate are scratched with a stainless steel knife, and then placed in a screw-capped tube. Experimental blank is set by scratching equivalent amount of silica gel from blank area of the plate into another tube. A standard curve is constructed by measuring the absorbance of standard samples. Test samples are quantitated by comparing the absorbance with the standard curve or regression analysis according to the formula: Total saponin content (mg/ml) = C (total saponins in μg)/2 × V (sample volume, μl).

(b) Methylophiopogonone A/B (Zhu *et al.* 1988b): (i) preparation of sample and standard solutions – Shengmai tonic (10 ml) is extracted five times, in a counter-current manner, each with 10 ml ether in five 50 ml-separating funnels. The ethereal layer from each separating funnel is washed twice with 5 ml of water, and each washing is done with the same ethereal washing (10 ml). The six ethereal extracts are combined, dried at reduced pressure, and the residue reconstituted in 2 ml methanol, serving as the test sample. To prepare the standard solution, 1.65 mg methylophiopogonone A and 2.18 mg methylophiopogonone B in 2 ml methanol, respectively. An aliquot (30 μl) of this solution is then made up to 1ml with methanol in a volumetric flask and serves as the mixed-standard solution. (ii) HPLC Analysis – mobile phase: methanol-water-acetonitrile (64:38:26); column: $C_{18}$; flow rate: 1.0 mm/min; sensitivity 8; detection wavelength 297 nm; microprocessing parameters: WIDTH 30, SLOPE 200, DRIFT 0, MINOR 100, T-DBL 0, LOCK 10, STPTM 25, AT-TEN 3, SPEED 2.5, METHOD 41, SPLWT 100, ISWT 1. The sample volumes of the test sample and standard are 10 μl and 5 μl respectively, and the amount is quantitated by comparing the standard and test values.

(c) Schisandrins (Zhu *et al.* 1988a): (i) sample extraction – Shengmai tonic (10 ml) is extracted five times, in a counter-current manner, each with 10 ml ether in five

50 ml-separating funnels. The ethereal layer from each separating funnel is washed twice with 5 ml water, and each washing is done with the same ethereal washing (10 ml). The six ethereal extracts are combined, dried at reduced pressure, and stored for future use. (ii) Preparation of sample and standard solutions – the residue of the above ethereal extract is reconstituted in 5 ml methanol. An aliquot (1 ml) of the methanol solution is then added to a 2 ml volumetric flask, combined with 100 μl of an internal standard solution of naphthalane (3 mg/2 ml methanol) and 50 μl internal standard, made up to 2 ml with methanol, serving as the standard solution. (iii) HPLC Analysis – mobile phase 73 per cent methanol; column: $C_{18}$; flow rate 1.0 ml/min; column pressure 2000 kg/cm; detection wavelength 254 nm; sensitivity 8; microprocessing parameters: WIDTH 30, SLOPE 200, DRIFT 0, MINOR 100, T-DBL 0, LOCK 4, STPTM 1000, AT-TEN 3, SPEED 2, METHOD 41, SPLWT 100, ISWT 1 and a recorder speed of 10 mm/min. The sample volume for both test sample and standard is 3 μl, and the amount is quantitated by comparing the standard and test values.

## III. Manufacturers

There are currently 124 manufactures producing different dosage forms of Shengmai San (Zhang 1997).

## REFERENCES


Chang, J. *et al.* (1998) Studies on the effect of natural flocculant on the fine processing of Shengmai San extract, *China Journal of Traditional Chinese Medicine*, **13**(2), 22.

Chen, Q. (1998) *Pharmacological and clinical basis of renowned traditional Chinese patent medicines*, People's Medical Publishing House, China, p.382.

Hu, C.L. *et al.* (1998) Clinical applications of Shengmai San and its injection, *Traditional Chinese Patent Medicine*, **20**(12), 34.

Liu, H.Q. *et al.* (1996) Studies on the application of ultrafiltration techniques in the production of Shengmai San tonic, *Chinese Traditional and Herbal Drugs*, **27**(4), 209.

Ministry of Health, P.R.China (1993) *Traditional Chinese medicine and patent medicine preparations*, 7, 47.

Ministry of Health, P.R.China (1995) *Traditional Chinese medicine and patent medicine preparations*, 10, 41.

Ministry of Health, P.R.China (1996a) *Traditional Chinese medicine and patent medicine preparations*, 11, 47.

Ministry of Health, P.R.China (1996b) *Traditional Chinese medicine and patent medicine preparations*, 11, 46.

Ministry of Health, P.R.China (1996c), *Traditional Chinese medicine and patent medicine preparations*, 11, 45.

Ministry of Health, P.R.China (1997a), *Traditional Chinese medicine and patent medicine preparations*, 12, 39.

Ministry of Health, P.R.China (1997b), *Traditional Chinese medicine and patent medicine preparations*, 15, 50.

Ministry of Health, P.R.China (2000), *Chinese Pharmacopoeia*, 436.

Shao, Y.S. (1999) Methods of removing turbidity in Shengmai San tonic, *LiShiZhen Medicine and Materia Medica*, **10**(1), 31.

Xie, M. (1990) *Zhongyi Fangji Xiandai Yanjiu {Current research in traditional Chinese medicinal prescriptions}*, Xue Yuan Press, China, 554.

Yan, K.D. and Jiang, X.R. (1990) Research on the quality control of Shengmai San tonic. In Y.Q. Yan (ed.) *Integrated Study of Shengmai San Tonic*, CMPSP, China, pp. 50–97.

Zhang, J.K. (1997) *Yiyao Gongzuozhe Yewu Zhishi Shouce {Handbook for Healthcare Workers}*, China National Pharmaceutical Industry Corporation, Beijing, 514.

Zhang, M.H. (1994) Comments on the possible improvement in the manufacturing process of Shengmai San set out in the Chinese Pharmacopoeia, *China Journal of Chinese Materia Medica*, 19(4), 207.

Zhu, C.X. *et al.* (1988a) Studies on the chemistry of Shengmai San (SMS) preparations (I) – determination of active ingredients in Fructus Schisandrae by TLC densitometry, *Yaoxue Fenxi Zazhi*, 8(2), 71.

Zhu, C.X. *et al.* (1988b) Studies on the chemistry of Shengmai San (SMS) preparations (II) – determination of isoflavanoids in Radix Ophiopogonis by reversed phase HPLC, *Yaoxue Fenxi Zazhi*, 8(6), 343.

Zhu, C.X. *et al.* (1989) Studies on the chemistry of Shengmai San (SMS) preparations (III) – determination of saponins in Radix Ginseng by TLC, *Yaoxue Fenxi Zazhi*, 9(1), 5.

# Glossary

| | | |
|---|---|---|
| *Assistant* | 佐 | *See Monarch*, *Minister*, *Assistant* and *Guide* |
| *Asthenia* | 虛證 | A syndrome produced by the deficiency of *vital energy* and *essence*, lowered resistance and hypofunction of the body, commonly seen in the individual with general debility and the patient suffering from a long-standing illness; manifested as lusterless complexion, listlessness, shortness of breath, weak voice, palpitation, insomnia, poor appetite, thick and tender tongue, feeble pulse, etc. |
| *Blood* | 血 | An important component of the body derived from the refined substance of the food through a series of complex processes. It circulates in the *blood* vessels to nourish all parts of the body. |
| *Blood stasis* | 血瘀 | A morbid condition with stagnation of *blood* circulation or coagulation of the extravastated *blood*; mostly due to stagnation of *vital energy*, deficiency of *vital energy* and *blood*, *blood-heat* or trauma. |
| Channels | 經脈 | The main passage connecting different parts of the body in which the *vital energy* and *blood* circulate. *See also* Meridian. |
| *Chest bi-syndrome* | 胸痹 | A disorder due to the accumulation of *Yin-evils* (*phlegm-wetness*, *blood stasis*, etc.) leading to the hypofunction of the *chest-Yang*, the stagnation of *vital energy* and the obstruction of *collaterals*; manifested as feeling of oppression over the chest, chest pain radiating to the back, shortness of breath, inability to lie flat, etc. |
| *Cold* | 寒 | 1) Referring to the *cold-evil*, one of the *six evils*, and is *Yin* in nature, usually damages *Yang*-energy and affects the activity of *vital energy* and *blood*. It may cause the symptoms such as chillness, fever, headache, general aching, abdominal pain, diarrhea, etc. 2) Diseases characterized by the hypofunction of the body. |

| | | |
|---|---|---|
| *Cold and heat* | 寒熱 | 1) Two principal syndromes for differential diagnosis serving as the concrete manifestations of the changes of *Yin* and *Yang*, and considered as a significant reference for the treatment of disease. 2) Referring to chilliness and fever. |
| *Collaterals* | 絡脈 | *See* Meridian |
| *Deficiency* | 虛證 | *See Asthenia* |
| *Depletion* | 脱證 | Also Collapse-syndrome. A syndrome due to exhaustion of *vital energy*, *blood*, *Yin* and *Yang* and the failure of important organs, manifested as profuse sweating, deadly *cold* limbs, incontinence of urine and feces, listlessness and fading pulse or even coma. |
| *Treatment by Differentiation of Signs and Symptoms* | 辨證論治 | Making a diagnosis and selecting the treatment based on the analysis and comprehension of the clinical data collected by the four methods of examination with the basic theories of *viscera*, meridian and pathogenesis. |
| *Dry; dryness* | 燥 | 1) Referring to the *dryness-evil*, one of the *six evils*, and tends to consume the *body fluid*, and may cause symptoms as conjunctival congestion, dry mouth and nose, non-productive cough, hypochondriac pain, constipation, etc. 2) Referring to *dryness-syndrome* due to consumption of *Yin-fluid*. |
| *Eight Principal Syndromes* | 八綱 | The eight aspects for differential diagnosis, i.e., *Yin* and *Yang*, *superficies* and *interior*, *cold* and *heat*, *asthenia* and *sthenia*. |
| *Eight Therapeutic Principles* | 八法 | Also: eight medical treatment; including *diaphoresis*, *inducing emesis*, *purgation*, *reconciliation* (regulation of the functional relationship between the internal organs), *warming*, *heat-clearing*, *resolution* and *invigoration*. |
| *Guide* | 使 | *See Monarch*, *Minister*, *Assistant* and *Guide* |
| *Heart* | 心 | One of the five solid *viscerae* and the most important one of the body. Its main function is to maintain the normal *blood* circulation. In TCM, the *heart* is considered to be closely related to the functional activities of the higher nervous system. The pathological changes of the *heart* are reflected in the disturbance of the *blood* circulation and the activities of higher nervous system, especially in that of the mental and emotional activities. Moreover, the secretion of sweat glands and the changes of the tongue are also related to the *heart*. |
| *Heart*-clearing | 清心 | A treatment for clearing away the *heat-evil* in the pericardium out of the body by the application of |

| | | |
|---|---|---|
| | | drugs with actions of clearing *heart-heat*, cooling *blood* and nourishing *Yin-fluid*. |
| *Heart-Yang* | 心陽 | The *Yang-energy* of the *heart*. In case of deficiency of *heart-Yang*, *cold syndrome* such as *coldness* of the limbs, feeble and impalpable pulse may appear. |
| *Heart-Yin* | 心陰 | The nutritious fluid of the *heart*, which is a component of *blood*, bearing a close relation to the *heart-blood* physiologically and pathologically, and also to the condition of *lung-Yin* and *kidney Yin*. |
| *Heat* | 熱 | 1) Referring to *heat-evil*. 2) One of the *eight principal syndromes*, a syndrome of hyperfunction of *Yang-energy*, manifested as *heat-syndrome* such as fever, congested conjunctiva, flushed face and thirst, etc. 3) A therapeutic method, i.e., the therapy by *warming* or the therapy by *expelling cold*. 4) One of the four characters of Chinese medicinal herbs. |
| *Heat-evil* | 熱邪 | The evil that causes symptoms of *Yang* and *heat* in nature, such as fever, noisy breathing, local redness, swelling, *heat* and pain, constipation, etc. |
| *Interior* | 裏 | *See Superficies and interior* |
| *Jue*-syndrome | 厥證 | 1) A temporary suspension of consciousness. 2) *Coldness* of the extremities. 3) The critical case of dysuria. |
| *Liver-Qi* | 肝氣 | Also *liver-energy*, referring to the function of the *liver*, including certain functions of nervous, digestive and endocrine systems. |
| *Lung* | 肺 | One of the five *viscerae*. Its main functions are to control respiration, to regulate fluid metabolism and to help the *heart* in regulating *blood* circulation, so as to provide the substances necessary for the physiological activities of the organs of the body. Furthermore, the *lung* is closely related to the body resistance to pathogens. |
| *Lung-Qi* | 肺氣 | Referring to the functional activities of the *lung*. |
| Meridian | 經絡 | An important component of the human body including the *channels* and their *collaterals*. It acts as the important route for circulating *vital energy* and *blood*, connecting *viscerae* with extremities, communicating the upper with the lower and the *interior* with the *superficies*, and regulating the activities of *viscerae* and other parts of the body, and plays an important role in joining the tissues and organs of the body to build up an organic entity. |

| | | |
|---|---|---|
| *Minister* | 臣 | See *Monarch*, *Minister*, *Assistant* and *Guide* |
| *Monarch, Minister, Assistant and Guide* | 君、臣、佐、使 | The terms signify the different effects of the drugs composed of a prescription. The *Monarch* acts as the chief drug for treating the disease and is composed by one or more drugs; the *Minister* serves to intensify the effect of the *Monarch*; the *Assistant* helps to deal with the secondary symptoms or inhibits the potent effect or toxicity of the *Monarch*; and the *Guide* leads the other drugs to the diseased part and balances the effects of the drugs. |
| *Nature* | 四氣 | Also the *Four Characters* – referring to the property of Chinese medicinal herbs concerning with its therapeutic effect, i.e., *cold*, *hot*, *warm*, and *cool*. |
| *Phlegm-dampness* | | See *Phlegm-wetness* |
| *Phlegm-wetness* | 痰濕 | A pathogenic factor produced by prolonged retention of *evil wetness* in the body as a result of the hypofunction of the *spleen*. It may cause a syndrome manifested as paroxysmal and productive cough with thin, whitish sputum, feeling of oppression over the chest, nausea, poor appetite, enlarged tongue with smooth and greasy fur, etc. |
| *Qi* | 氣 | 1) Air breathing in and out of the body during respiration. 2) Refined and nutritious substances flowing in the body. 3) *Vital energy* – the functions of various organs and tissues of the body. |
| *Qigong-respiratory exercise* | 氣功 | Also *Breathing exercise* – A mental and physical self-training for the prevention and treatment of diseases, and also for health care and prolongation of life, by which life activities are self-adjusted and self-controlled with the help of the inducement of mind and the regulation of respiration and spirit. |
| *Sour* | 酸 | *Sour* drugs of sour taste which possess the action of astringing or antidiarrhea. |
| *Spleen* | 脾 | One of the five *viscerae*, which, in TCM, does not completely match the organ designated in western medicine from the standpoint of structure, location and function. It has the functions of digesting food, absorbing and transporting nutrients to the body tissues. The *spleen* also serves to control the *blood* and to keep the *blood* circulating within the vessels, and takes part in the regulation of fluid metabolism. Some diseases of the digestive system, edema and chronic hemorrhagic diseases are usually attributable to malfunction of the *spleen*. |

*Sthenia* 實證 A syndrome occurring in a relatively strong patient exposed to virulent evil or dysfunction of the *viscera* leading to the stagnation of *vital energy* and *blood*, *phlegm-retention*, indigestion, etc.; manifested as high fever, thirst, irritability, delirium, abdominal distention, pain and tenderness, constipation, oliguria with deep-colored urine, red tongue with rough, dry and yellowish fur, solid and strong pulse, etc.

*Superficies and interior* 表裏 Two principal aspects for estimating the site and severity of a disease. In general, a disease involving the *skin*, *hair* and *meridian* is considered as a mild form of impairment in the *superficies*, while those involving the *viscera* as a more serious form of impairment in the *interior*.

Taste 五味 *Five tastes* – referring to the properties of Chinese herbs, i.e., acrid, sweet, sour, bitter and salty.

*Viscerae and organs* 臟腑 The five *viscerae* (*Zang-organs*) – a collective-name for the *heart*, *liver*, *spleen*, *lung* and *kidney*. While each of them is considered as a functional unit, the terms for the five *viscerae* in TCM do not completely match those used in western medicine. The six *organs* (*Fu-organs*) – referring to the *gall bladder*, *stomach*, *small intestine*, *large intestine*, *urinary bladder* and *tri-Jiao*.

*Vital energy* See *Qi*

*Water and dampness (disease)* 濕病 Disease due to *wetness-evil* – manifested as heavy feeling of the body, soreness of limbs, poor appetite, diarrhea, abdominal flatulence, even edema of the face and limbs.

*Wei-Qi* 衛氣 A kind of *Yang-energy* in the human body, derived from the digestion and absorption of foods by the *spleen* and *stomach*. Its functions include protection of the skin and muscle, resistance against the *exogenous evils* and regulation of the secretion of sweat.

*Wind cold* 風寒 A type of common *cold* due to the attack of exogenous *wind-cold evil*; manifested as fever, chilliness, headache, anhidrosis, stuffy nose with watery discharge, sneezing, cough with irritation of throat, arthralgia, thirstlessness, thin and whitish fur on the tongue, floating and tense pulse, etc.

*Yang* 陽 A philosophical term in ancient China, referring to the things or characters opposite to *Yin*. The condition which appears as active, external, upward, hot, bright, functional, exciting and hyperactive, is attributive to *Yang*. In TCM, it is widely used for explaining

| | | |
|---|---|---|
| | | the physiological and pathological phenomena of the human body, and for guiding the diagnosis and treatment of diseases, for example, *superficies-syndrome*, *heat-syndrome*, *sthenia-syndrome* are ascribed to *Yang*. |
| *Yang deficiency* | 陽虛 | A *cold-syndrome* in the *interior* resulting from the insufficiency of *Yang*-energy; manifested as fatigue, shortness of breath, intolerance of *cold*, *cold* extremities, spontaneous perspiration, pallor, polyuria with watery urine, diarrhea, pale and tender tongue, feeble and large or sunken and small pulse. |
| *Yang-hyperactive* | 陽亢 | A morbid condition with hyperactivity of *Yang-heat*, indicative of *sthenia* of *heat-evil* while the healthy energy of the patient is vigorous; manifested as high fever, no sweating, noisy breathing, irritability and thirst. |
| *Yang-Qi* | 陽氣 | Also *Yang-energy*. It is the opposite of *Yin-energy*. As for the function and morphology, *Yang*-energy stands for function. As for the physiological and pathological phenomena, *Yang-energy* stands for those which are external, upward, hyperfunctioning, reinforcing, light, etc. |
| *Yin* | 陰 | A philosophical term in ancient China, referring to the things or characters opposite to *Yang*. The condition which appears as inert, internal, downward, *cold*, dim, material, inhibitive and declining is attributive to *Yin*. In TCM, it is widely used for explaining the physiological and pathological phenomena of the human body, and for directing the diagnosis and treatment of diseases, for example, *interior-syndrome*, *cold-syndrome*, *asthenia-syndrome* are ascribed to *Yin*. |
| *Yin*-fluid | 陰液 | The components of *body fluid* belonging to *Yin*, such as *essence*, *blood*, thin and thick fluid, etc. |
| *Ying* | 營 | Nutrition – one of the essential substances for body life activities. It is derived from the digested food and absorbed by the internal organs, and circulates through the vessels to nourish all parts of the body. |
| *Ying-Qi* | 營氣 | Essential substance circulating in the vessels, responsible for the production of *blood* and nourishment of body tissues. |

# Index